AF880241

This page intentionally left blank

The Asian Ratsnakes and Kin of Greater China

中华锦蛇

---

by Kevin R. Messenger, Ph.D.

Nanjing Forestry University

With contributions by Matthew G. Most and Zang Jing (臧晶)

and illustrations by Adeline Robinson.

Translations provided by Zang Jing (臧晶), Luo Tingkun (骆廷昆), Zhu Xinxin (朱欣欣),

Yang Yi (杨毅), Zhu Yufei (朱聿霏) and Alden Zhang (张硕).

凯文·梅辛杰

南京林业大学

KDP Publishing

2021

Kevin R. Messenger 凯文・梅辛杰
The Asian Ratsnakes and kin
of Greater China

Front cover:
Elaphe anomala on the Great Wall of China, K. Messenger 凯文・梅辛杰
Back cover (clockwise from upper-left): *Elaphe taeniura friesi, Euprepiophis mandarinus, Gonyosoma prasinum* (Parinya Pawangkhanat), *Elaphe anomala, Oreocryptophis porphyraceus vaillanti, Elaphe moellendorffi, Elaphe carinata,* and *Archelaphe bella chapaensis* (Matthew Most).

First Printing: 2021
ISBN 978-1-7363177-0-9

Publisher: Kevin Messenger 凯文・梅辛杰
Nanjing Forestry University
159 Longpan Road
Nanjing, Jiangsu, 210037
Email: kevinrmessenger@gmail.com
Website: youtube.com/c/kevinmessenger

Layout: Kevin Messenger 凯文・梅辛杰

Proofreading: W. Scott Lupien, Zang Jing (臧晶), Zhu Xinxin (朱欣欣), Luo Tingkun (骆廷昆)

Printing: KDP
Binding: KDP

ISBN 978-1-7363177-0-9

Dedication

To my parents, Robert and Karen, who always encouraged me, fostered my love for herps, and were supportive of whatever avenues I would take with regard to my professional career.

献词

我对两栖和爬行动物的热爱源于父母的鼓励与培养，

能将这份热爱转化为事业更离不开他们所给予的莫大支持，

谨以此书献给我的父母—— Robert Messenger、Karen Messenger。

Contents / 目录

Acknowledgements / 致谢

Most people that know me, know that I am usually pretty photo heavy in my writing. So for the general reader that does not know me, I apologize for the lengthy acknowledgements. In addition to the fantastic contributing photographers, many of whom I now call friends, several other friends and colleagues helped to make this project possible. For them, I am now, and forever will be, incredibly grateful.

此次书籍得以顺利出版离不开为本书提供图片的摄影师们，更离不开诸多同事和朋友的帮助。对此，我将永怀感激之情。

Dr. Harold Heatwole is my mentor from my undergrad years at North Carolina State University. He was instrumental in enabling me to embark on this journey into China's herpetology and the unbelievable experiences I have enjoyed there over the past 14 years. It was with his recommendation that I applied for a herp survey opportunity in central China. At the same time, I was still an undergraduate student at North Carolina State University. The opportunity was offered by Dr. Craig Stanford of the University of Southern California, in collaboration with Dr. Yiming Li of the Chinese Academy of Sciences in Beijing. My application was accepted, and thus began my interest in the natural history and herpetofauna of China.

Harold Heatwole博士是我在北卡罗来纳州立大学本科学习期间的导师。在他的帮助下，我对中国两栖和爬行动物的研究之路得以拉开序幕。14年前，经 Harold Heatwole博士的推荐，我以本科生的身份申请了一项在中国中部地区开展的两栖和爬行动物调研工作。这一工作机会源于南加利福尼亚大学Craig Stanford博士与中国科学院李义明博士的一个合作项目。有幸被录取之后，我对中国自然史和两栖爬行动物的浓厚兴趣就此萌芽，一段无比美妙的华夏之旅也就此开启。

This first trip, in the Shennongjia National Park, in western Hubei, also had a transformative impact on me. The rangers and officers in Shennongjia that helped me navigate my first trip to China, the friends I made along the way, and the incredible scenery and cultural experiences I gained all helped shape my future. For this, I am extremely grateful to all the staff and officers in Shennongjia. It is still among my favorite locations in China.

那段旅行，我来到位于湖北省西部的神农架国家公园。指引我在林间穿行的工作人员，一路结交的伙伴，无与伦比的景色以及曼妙的异国文化风情，无不对我的未来之路产生了变革性的影响。为此，我想向神农架国家公园的全体工作人员表达我的感激之情。时至今日，那里仍是我在中国最向往的热土之一。

I should acknowledge and thank my parents, Bob and Karen, for all of their support over the years. Both of them visited China with me in 2008 for a month-long herping and Chinese culture expedition in the Shennongjia National Park, located in western Hubei province.

我也要特别感谢父母多年来对我的支持。他们于2008年随我一起到访中国，同样在神农架国家公园度过了为期一个月的美好时光。

My story as an American herpetologist in China has been most directly facilitated by my Ph.D. professor from Alabama A&M University, Dr. Yong Wang. With his aid, I traveled to China every year from 2011 to 2016 to work on my Ph.D. dissertation. Dr. Wang continues to be a trusted advisor and resource for all things China.

作为一名来自美国的两栖和爬行动物学家，我的事业得以在中国起步首先要感谢王勇教授，他是我在阿拉巴马农业机械大学攻读博士学位时的导师。2011至2016年间，在他的协助之下，我每年都能前往中国开展我的博士课题研究。时至今日，王勇博士仍是我在中国最值得信赖的良师益友之一。

For me, the most influential mentors and colleagues in China would be Dr. Yulong Ding and Dr. Honghua Ruan, both of Nanjing Forestry University (NFU), in Nanjing, Jiangsu province. Dr. Ding recommended that I enroll at NFU, which I did in 2014. Following graduation in 2017, Dr. Ruan offered me a faculty position, which I currently hold at the time of this writing.

在中国，对我影响最深远的导师和同事当属南京林业大学的丁雨龙博士和阮宏华博士。2014年，丁雨龙博士引荐我进入南京林业大学继续学习深造，而阮宏华博士则使我在2017年毕业后有幸成为学校教师团队的一员。撰写本书时，我仍任教于南京林业大学。

Mr. W. Scott Lupien is an American businessman living and working in China. He is an avid and knowledgeable herpetologist, and fortunately for me, is fluent in Mandarin, and generous in his willingness to use it to my benefit. I met Scott online through the Field Herp Forum. Scott read posts of my China adventures on FHF and reached out to me, inviting me to visit him the next time I came to China. I met Scott face to face on my second trip to China in 2008, and ever since then, he has given me support, friendship, and counsel, which ultimately shaped many aspects of my career.

卢彬先生是一位狂热而又博学的两栖和爬行动物学家。他来自美国，在北京定居、经商已十年有余。值得庆幸的是，他也十分精通普通话，这点让我受益匪浅。我们的相识始于互联网论坛“Field Herp Forum”。在该论坛上，我会定期发布、更新我在中国的游历见闻，也因此与同为论坛用户的卢彬结缘。2008年，当我第二次到访中国时，卢彬便邀我在线下一聚。从那时起，卢彬给予我的支持和建议，彼此间的友谊，皆对我的职业生涯产生了积极的影响。

In the field, I've been fortunate to meet and befriend many other amazing people: Recognition and gratitude are owed to Dan Rosenberg, Kevin Caldwell, Bill Murphy, Noel Thomas, Will Sargent, Abdelhamid Bizid, Dong Xie, Wang Ming, Zong Xu, Zhang Fang, Rob Ferguson, Jack Ferguson, Eddie Yam, Hollis Dahn, Justin Waraniak, Iwo Gross, Tim Bova, Thomas Reed, Ringo Wong, Bosco Chan, and Tommy Swift. Thank you to Matthew Most and Rainer Fesser, who helped me fill out some of the areas of the captive sections which is certainly my weakest area of expertise. And I owe significant thanks to Zang Jing, Yang Yi, Zhu Yufei, Zhu Xinxin, Luo Tingkun, and Alden Zhang for their immense help in translating this text into Chinese. Thanks to Wang Xiaohe and Shi Jingsong for additional reviews on some of the species entries. Additional thanks to my dad and Scott Lupien for additional edits to the English text. I need to thank those individuals who helped me with distribution data outside of China, in order to make the distributions more complete, to the benefit of all. For this help, I'd like to thank Chaitanya Shukla, HT Lalremsanga, Parinya Pawangkhanat, and Peter Brakels, especially for the countries of India, Thailand, Vietnam, and Laos, respectively. Many others who helped me arrive at this point in my professional life are not mentioned by name here because the list is too long. To all of you, your friendship and support will be forever valued.

游走在中国的山川大河、乡野小径，我也很庆幸身旁始终有一群好友相扶相伴，我想向他们致以我的谢意，他们是Dan Rosenberg、Kevin Caldwell、Bill Murphy、Noel Thomas、Will Sargent、Abdelhamid Bizid、谢东、王敏、周宗旭、张方、Rob Ferguson、Jack Ferguson、Eddie Yam、Hollis Dahn、Justin Waraniak、Iwo Gross、Tim Bova、Thomas Reed、黄光智、陳葦樂博士和Tommy Swift。感谢Matthew Most和Rainer Fesser在蛇的人工饲养上对本书的贡献，这是我知识面相对欠缺的环节。同时，我也想向本书的中文翻译团队表达感激之情，他们是臧晶、杨毅、朱聿霏、朱欣欣、骆廷昆和张硕。感谢王小荷博士和史静耸博士为本书提出的宝贵意见。卢彬以及我的父亲为本书英文部分的编辑提供了巨大的帮助，同样向他们致以谢意。在地理分布版块，能准确获悉某些物种在中国以外如印度、泰国、越南和老挝等国的分布情况离不开多位朋友的帮助。为此，我想向Chaitanya Shukla、HT Lalremsanga、Parinya Pawangkhanat和Peter Brakels致以我的谢意。最后我还想感谢其他许许多多在事业上给予我帮助的朋友，请原谅我无法在此一一例举，你们的友谊与支持将永存心间。

Foreword / 推荐序言

Recent years have seen an ever-increasing surge in the publication of high-quality books and monographs on a wide variety of taxa of Chinese animals. These range in linguistic scope from those solely in English to those in English with abstracts in Chinese, to those in Chinese with English summaries, to those only in Chinese, with an overall shift toward greater reliance upon Chinese. This array of scientific literature reflects the rapid growth and development of scientific research in China and its role in enhancing China's global impact on science.

近年来，关于中国各类动物的优质出版物越来越多。从语言的角度来看，作品所使用的语言从完全为英语，到以英语为主但附有中文的摘要，到以中文为主但附有英文的摘要，到完全使用中文，总体趋势体现了对中文的愈发依赖。这些科学文献反映了中国在科学研究领域的快速发展，也体现了中国科研水平在全球的影响力正与日俱增。

Historically, scientific information was originally published in Greek or Latin, later to be superseded by French as the '*lingua franca,*' and finally by English as the universal language of science. When I was a post-graduate student in zoology in the late 1950s in the United States of America, all Ph.D. candidates at my university were required to write their dissertations in English, but they also had to demonstrate proficiency in French and German. Such foreign language requirements are no longer in place, and English is nearly universally required in academia worldwide.

从历史角度来看，科学信息最初是以希腊语或拉丁语出版，之后一度被法语所取代，最后英语成为全球科学界的通用语言。上世纪50年代后期，我在美国攻读"动物学"硕士学位。当时所有的博士生必须使用英语书写论文，同时还需精通法语和德语。如今，语言的要求不再那么严苛，全球的学术界已普遍采纳英语作为通用语言。

Reliance on one language for science has an advantage in that it allows direct communication between all highly trained scientists everywhere, but in biology has the disadvantage of hindering the transmission of scientific information to naturalists, teachers, and interested laymen in non-Anglophone countries. Publishing only in a language that is not the current universal one has the reverse effect, with better accessibility to citizens of the country of origin, but the results are not disseminated to scientists other than those versed in the local language. As an example, the findings of an ecological study of ants in Xinjiang, which I carried out jointly with Chinese collaborators, were published only in Chinese, without an English summary. I wrote the original draft in English, and then my co-authors translated it into Chinese. The ironic result was that I am unable to read a paper of which I am the senior author.

在科学领域，依赖某一门语言的优势在于它使世界各地所有训练有素的科学家得以直接交流。然而，在生物学领域，它可能会阻止科学信息的有效传递，毕竟很多博物学家、教师和业余爱好者的母语并非为英语。若以一门非当前通用语言进行出版则会产生完全相反的效果：该语言的使用人群能够轻易地获取信息，但缺乏相关语言背景的人士则会被排除在外。例如，我与中方合作者共同开展了一项对新疆地区蚂蚁的生态研究。研究结果仅以中文发表，未有英文总结。我以英文书写了初稿，我的合著者随后将其翻译成中文。颇具讽刺意味的是，作为论文的主要参与者，我最后居然无法阅读自己的文章。

The ideal solution would be to publish scientific books and journals both in Chinese and in English. However, I know of very few such dual-language books. In most cases, the cost of publication would be prohibitively high. It is conceivable that in the future, Chinese may become the universal language. That is not the case now, and unless, or until, that happens, which language to use for scientific reports remains a conundrum. The most workable solution would seem to be to publish in the language that is native to the authors but with an abstract that is as detailed as possible in the alternative language.

理想的对策是用中英双语对科学书籍和期刊进行出版。然而这类双语刊物非常罕见。大多数情况下，出版成本也会过高。可以设想，中文或在未来成为全球通用的语言。不过在此之前，科学报告以何种语言书写仍将是一大难题。最可行的解决方案似乎是以作者的母语进行发表，但摘要可用另一门语言书写并尽可能做到详实。

This book deserves to reach not only an international audience of specialists and hobbyists in herpetology, but also that segment of the Chinese public that includes biology teachers, naturalists, or merely people with a healthy, inquisitive interest in snakes. The Asian Ratsnakes and Kin of Greater China brings together a vast amount of scattered information not otherwise readily available in a concise form, as well as new data assembled by the author's own research. This work indicates the characteristics of the species of ratsnakes and provides a detailed description of each species. Accordingly, the Asian Ratsnakes and Kin of Greater China can serve as a field guide and a manual for identification. It summarizes what is known of the habits and natural histories of these snakes, thereby also revealing what isn't known; it should encourage interested amateurs to provide data to fill in those gaps in knowledge. The hobbyist who wants to keep ratsnakes in captivity also can benefit from the information provided.

除全球的两栖和爬行动物学家与业余爱好者外，中国的生物学教师、博物学家、甚至对蛇类感兴趣的社会大众也都将是本书的受众。在书中，作者对大量分散的信息进行了汇总，添加了通过自身研究所收集的新数据，并最终将它们以简明的形式呈现给读者。本书也充分展现了锦蛇的诸多特点，并对所涉及的每个物种都进行了详细的描述。因此，本作品可作为一本户外指南或辨识手册。作者归纳了书中蛇种在习性和自然史方面的各项已知信息，还由此揭示了大量的未知信息。本书也鼓励业余爱好者提供数据来弥补相关知识上的空白。有兴趣饲养锦蛇的爱好者也能从书中获得启发。

The known distribution of species, sometimes with maps, is provided and, again, should be a stimulus both for professionals and amateurs to contribute new data on localities extending the known ranges of these snakes. Finally, the photographs are excellent and mostly taken by the author, and consequently available nowhere else.

本书对各物种的介绍涵盖对其分布情况的描述，在地图上也做了相应的标注。这或将激励专业人士和业余爱好者在此方面贡献新的数据，从而扩展某些物种已知的分布范围。最后，书中的高质量照片大部分由作者本人所拍摄，可谓独家分享。

The appeal of this treatise is broader than reached only by an English version, which was one of the primary goals the author intended to surmount with this work. Further works on other taxa of snakes would be a valuable additional milestone in stimulating interest in snakes and in disseminating knowledge about them.

以中英双语出版无疑使本书更具吸引力，这也是作者的初衷之一。作者未来关于其它蛇类分类群的作品或将再一次激发公众对蛇类的兴趣并促进相关知识的传播。

When Kevin Messenger was a student of mine at North Carolina State University in the United States, I passed on to him an appeal for applicants to conduct research on reptiles in China. He applied and was successful. To prepare himself for that work, he enrolled in a university course in Mandarin and began eating at a local Chinese restaurant to hear Chinese being spoken and practice speaking with the restaurant's staff, all of which attests to his interest in the herpetology of China and his dedication as a scientist. Thus began the journey that eventually led to his receiving a doctorate from a Chinese university and later serving in China as a faculty member, and finally to producing the present publication.

本书作者凯文曾是我在美国北卡罗来纳州立大学任教时的一名学生。当时，一项在中国开展的爬行动物调研工作正在招募人选，我便将这一消息告知了凯文，他随即申请并成功入选。在启程前往中国之前，凯文做了大量的准备工作。他选修了汉语普通话课程，并在当地的中餐馆模拟演练自己的中文。所有这一切都体现了他对中国两栖和爬行动物学的浓厚兴趣以及作为科学家的献身精神。此后，凯文更是与中国结下了不解之缘：他在中国的大学获得了博士学位并在毕业后留校成为教师队伍的一员，他的第一本作品也成功地以中英双语进行出版。

Harold Heatwole,
University of New England,
新英格兰大学
Armidale, NSW 2350
阿米代尔，新南威尔士州
Australia
澳大利亚

Preface / 作者的话

The inspiration to write this book came out of a personal desire to do so, but also from suggestions from friends and colleagues to share in writing and photographs my unique perspectives accumulated from uncounted hours of research in the field. Like most herpetologists, while growing up, I collected books about reptiles and amphibians, and I was especially fond of field guides. Field guides were concisely written and were beautifully illustrated. Every time our family traveled to a new state in the USA, I would ask my parents to take me to a bookstore so that I could buy the "Reptiles and Amphibians of ______ (fill in whatever state)." My goal was to collect all the available field guides for snakes in the USA. And of course, I bought the regional guides as well. One of my favorite local field guides was the Peterson Field Guide to Reptiles and Amphibians of Eastern and Central North America. The quality of the plates was excellent, the maps were great, and the layout of the information was very user friendly. These (American) regional and state guides led me to have certain expectations when traveling to other countries. When I started traveling outside of the USA, I naively had the same expectations internationally as I previously had in the USA; I wanted to collect books on reptiles and amphibians in whatever country I was traveling to, and I thought I would be able to do so easily.

本书的写作灵感源于我的个人愿望以及好友、同事们的建议。同大多数两栖和爬行动物学家一样，在成长的过程中，我有收集两栖和爬行动物书籍的爱好。其中，户外指南类丛书令我情有独钟，因为这类书籍言简意赅，插图生动精美。年幼时每次到访美国其它州，我总会要求父母带我去当地的书店转转，这样我便有机会购买到介绍该州两栖和爬行动物的专业书刊。我曾一度计划将所能购买到的关于美国蛇类的户外指南刊物全部收入囊中。当然，我也会购买介绍某一地区的户外指南。时至今日，我觉得此类书刊的巅峰作品之一为*Peterson Field Guide to Reptiles and Amphibians of Eastern and Central North America*。该书插图优美、地图详实、信息的布局方式带给读者一流的阅览体验。这些关于美国各州及地区层面的指南书籍也令我每次出国都多了一份期待。因为我曾天真地认为，在我到访的国家，我能轻而易举就收集到有关当地两栖和爬行动物的书籍。

On my first trip to China, in 2006, I searched both online and at local bookstores for field guides, hoping for something like the Peterson guide. The choices were slim. Most of the books I found were written only in Chinese, aside from the scientific name, and used images which consisted of drawn renditions or photographs of dead museum specimens instead of living animals. The range distribution maps were quite vague, providing only approximate geographic regions of occurrence.

当2006年我首次来到中国，我立刻通过线上和线下渠道寻找与*Peterson Field Guide*系列旗鼓相当的作品，但选择面非常有限。中国的此类书籍中，除了物种的学名外，几乎所有信息都以中文书写。鉴别某一物种的图片很少使用活体照片，往往是以手绘图片或标本照片为主。对物种的分布描述也是相当含糊，仅提供大致的地理分布区域。

Although the basics of spoken Chinese are reasonably easy to pick up, learning how to read and write in the language is quite another matter for most, if not all, foreigners. I was surprised at this scarcity of choices in book resources, as there are frequently English guides in most other countries in the world. As much as I looked, I was unable to find any for China. The last quality English language publication on Chinese herps, from a field guide perspective, was Pope's 1935 classic: The Reptiles of China, which was going for around $500. In 1993, Zhao and Adler published: The Herpetology of China (in English), but this book was not so much a field guide as it was a collection of species, synonyms, a few photographs, and provincial occurrences. There was no information on each species' natural history or ways to identify them, which was my primary interest in buying a field guide. So, in the beginning, faced with limited resources and requiring quality, readable information, I created a proxy, a personal "reference" based on information I could pull out of my library. I made an amalgam of this data, printed off a copy, and used this throughout my first trip. It proved to be an invaluable resource.

尽管用汉语进行基本的口语交流尚容易应付，但进行阅读和书写对于大多数外籍人士来说难度可想而知。全球大多数国家都已有介绍自身特色的英文版指南刊物，但偌大的中国却并不在列，这点曾让我颇为吃惊。关于中国两栖爬行动物的此类书籍，最近的佳作还要追溯到Clifford Hillhouse Pope于1935年出版的经典——*The Reptiles of China*。该书售价在500美元左右。赵尔宓院士与美国鹰岩教授（Kraig Adler）合著的*The Herpetology of China*（中国两栖爬行动物学）一书在1993年以英文出版。该书涉及众多物种，虽然介绍了这些物种的省级分布以及同物异名并附有大量插图，却不能被归为户外指南类刊物。该书未涉及"物种的自然史"和"如何鉴别某一特定种"两方面信息，而这两点恰恰是户外指南类书籍中最能打动我的部分。面对有限的资源，出于对高质量可读信息的渴求，我当时决定基于自己多年来收集的信息亲手打造一本类似的参考书。而这本临时拼凑的笔记也伴我度过了在中国的首段探险历程。事后来看，它的确是一份无价之宝。

Left: The Reptiles of China, by Pope, 1935. Right: The Herpetology of China, by Zhao and Adler, 1993.

I upgraded the photographs in the proxy with my pictures as I progressed through the summer of 2006. Once I got back to the USA, with my newly acquired love for China and my desire to return there and conduct more fieldwork, I was continually looking for better field guides on reptiles and amphibians. My search was in vain because there were no such guides. I could buy books from the regions of China and could rely on pictures and scientific names. For natural history or identification information and maps, I would have to rely on non-Chinese sources.

2006年的盛夏一天天过去，原先笔记中的图片也一张一张地被我自己所拍摄的照片所替代。回到美国后，出于对中国的热爱以及对重回中国的憧憬，我开始长时间地物色更佳的户外指南类书籍。不过所有的寻觅都以失败而告终，因为压根就不存在这样的刊物。我会在中国的不同地区购买类似的书籍，其中的物种学名和物种图片对我仍具有很高的参考价值，但是对于物种的自然史、辨识依据以及分布情况，我则不得不依赖中国以外的信息来源。

With the online posting of my photographs from my China adventures, many fellow herpetologists suggested I write a book. I have finally decided to act on their suggestions. This book is created mostly from personal experience. Still, where such knowledge is lacking, I have relied on other experts and like-minded individuals to help fill in the gaps. Therefore, this book is also a summary and synthesis of information from various sources instead of original information.

随着我在网上陆续发布我在中国游历时所拍摄的照片，越来越多的朋友建议我将它们一并整理，然后以书本的形式进行出版。我也最终决定将之付诸行动。本书主要基于我的个人经历，对于个人阅历尚未涉及的部分，我将倚靠专家或朋友提供的可靠信息。因此，本书与其说是我个人的一手数据和信息，不如说是对多渠道信息的总结、提炼和融合。

The Asian Ratsnakes and Kin of Greater China will hopefully be the first in a series of English-friendly field books on Chinese and Asian reptiles and amphibians. It is my intention and hope to delve into other regions of China and gather useful data and images on these magnificent animals and present them to you in future volumes.

《中华锦蛇》有望成为我所创作的英语户外指南系列中第一本问世的作品。该系列将聚焦中国和亚洲的两栖爬行动物。接下来，我希望对中国其它地区展开深入的研究、收集更多有用的数据和图像、并将它们在之后的出版物中呈现给读者。

I hope all readers share the joy of this inaugural step for me as I dive into this endeavor.

本书的顺利出版意味着我已向此目标迈出了第一步，希望所有的读者都能够分享我的这份喜悦。

- Kevin Messenger, 5 October 2020

- 凯文·梅辛杰，2020年10月05日

Introduction / 简介

This book is written to accomplish two goals. The first goal is to provide an English language field guide and reference book for students, naturalists, hobbyists, and professional herpetologists who, like me, cannot read written Chinese. It is meant to be used as a resource for non-native dedicated reptile enthusiasts, like me, living, working, or visiting Greater China (mainland China, Hong Kong, Taiwan, and Macau). Since this book is written in both English and Chinese languages, not only could it be useful to the host population living in this beautiful country, but it could be used to break through language barriers; anyone unfamiliar with either language could point to a section of text to communicate. I ran into this dilemma many times in my early years researching snake populations in China. I often wished to describe various target species' specific habits or habitats to my Chinese colleagues who knew the land and its fauna well. If I had a book like this one, I could have pointed to the section that described these specifics, and my issues would have been resolved.

本书旨在实现以下两个目标。首要目的即为像我（作者）一样无法阅读中文的学生、博物学家、业余爱好者与专业两栖和爬行动物学家提供一本英语版的户外指南与参考书。在中国（含港、澳、台）生活、工作、访问的外籍发烧友皆可使用本书。本书由中英双语编著，广大的汉语读者亦是受众人群。通过点读本书中的某一段落，英文读者与汉语人群能完成基本的信息交流，从而在一定程度上打破不同语言间的壁垒，而这正是我早年间在中国研究蛇类时所遭遇的窘境。我常想向了解中国这片土地和其动物群的中国同事介绍某些物种的特殊习性或栖息地。如果我的手边能有一本类似的书籍，只要指向书中某一段落，那么很多问题便能迎刃而解。

Several domestic references currently exist, many published by the Chinese Academy of Sciences and various universities. However, they are all written in Chinese. As a result, for anyone who cannot read Chinese, they are nearly useless, except for the pictures: these books may still have value as photographic guides for foreign investigators who cannot read Chinese. For me, these books were helpful because they provided a picture of an animal, its scientific name, and usually, but not always, a geographic range in which one could find the animal. An excellent example of such a book that I used is "The Coloured Atlas of Reptiles of Sichuan Province," available from the Chengdu Institute of Biology. In The Asian Ratsnakes and Kin of Greater China, I hope to remedy this language barrier. My first goal is to provide a bilingual reference book that will permit China's herpetology to be accessed by a much larger number of like-minded people, providing renewed interest in the herpetofauna that lives in this extraordinary land with its ancient culture.

中国国内目前存在一些参考作品，其中许多是由中国科学院或各高等学府所出版，但全部只有中文版。对于无法使用汉语进行阅读的人群来说，仅书中的图片、物种学名、分布区域等版块具有参考价值。我时常翻阅的《四川爬行类原色图鉴》就是这样一个例子。该书由中国科学院成都生物研究所编写，书中各物种的学名、图片和地理分布还是令我有所受益。本书旨在弥补因语言不通而造成的遗憾。因此，我的第一个目的就是通过此双语参考书令更多志趣相投的朋友能够接触到中国的两栖和爬行动物学，从而激发大家对这片神奇而又古老土地上的两栖和爬行动物产生新的兴趣。

This book's second goal is to expand and improve upon the information currently known for the Asian ratsnakes included within this book. It will provide a detailed account of their natural history, a morphological description, an update on their respective geographical distributions, and some information on their captive husbandry.

第二个目标是扩展和优化本书所涉及的各锦蛇物种的已知信息，包括详细介绍它们的自然史、外观形态和人工饲养信息，并对各自的地理分布做出更新。

Notes on taxonomy / 分类注释

The taxonomy of many genera and species is unsettled worldwide. However, the situation is exacerbated in Asia right now as many Asian species are still poorly understood, and "ratsnakes" are among them. One of the issues with common names is that the words do not necessarily indicate relationships as they suggest other animal features or characteristics. If people see a snake frequently eating rats as its main diet, species will often be given a common name of "ratsnake" if there are no better descriptors that could be assigned to that animal.

从生物分类学的角度来看，世界范围内，许多种属的分类目前都尚无定论。这一情况在亚洲更是尤为明显。究其原因，学界对亚洲的部分物种仍知之甚少，其中就包括锦蛇（注：英文名为ratsnake，直译为"鼠蛇"）。常用名的一大问题是其暗示了某一物种的特征及特点，但却无法反映物种之间的亲缘关系。如果人们看到某一蛇种主要以鼠为食，那这一蛇种就经常被冠以"鼠蛇"之名，尤其是在没有更贴切的描述来为该物种命名的情况下。

This book will cover those classically assigned species into the genus *Elaphe* and their Kin, also known as the "ratsnakes." For convenience, several species that previously belonged to *Elaphe* and separated or split from this genus are also included, as they traditionally have been accepted as ratsnakes by scientists and hobbyists. Besides being closely related to each other, *Elaphe* and their kin also share many similar natural history traits that separate them from other Asian genera with the common name "ratsnakes." These traits include using constriction before eating their prey and often have an arboreal or partially arboreal lifestyle. Many other colubrid genera in Asia also have the common name "ratsnake," for example, the Indo-Chinese Ratsnake and Common Ratsnake (*Ptyas korros* and *P. mucosus*, respectively). As these snakes were never in the genus *Elaphe* and do not have similar habits aside from a preference for eating rodents, they are not included in this work.

本书旨在介绍锦蛇属（*Elaphe*）各成员及其近缘种。方便起见，此前归隶锦蛇属但之后改隶它属或另立新属的数个物种也会包含在内。因为从传统上来看，它们已被科学家和爱好者接受为锦蛇。除了亲缘关系以外，锦蛇及其近缘种在自然史方面也有许多共同点，这些特征将它们同亚洲其它以"鼠蛇（*ratsnake*）"为常用名的异属蛇种相区分。此类特征包括在进食前绞杀猎物，以及具有树栖或半树栖的生活方式。亚洲许多游蛇科的物种也以"鼠蛇（*ratsnake*）"为常用名，比如灰鼠蛇（英文名：Indo-Chinese Ratsnake / 学名：*Ptyas korros*）和滑鼠蛇（英文名：Common Ratsnake / 学名：*Ptyas mucosus*）。考虑到这些蛇种从未归隶锦蛇属，且除喜食啮齿类外，与锦蛇在习性上也没有其它的共同点，故本书将不做讨论。

The genera covered in this book include *Elaphe, Archelaphe, Coelognathus, Euprepiophis, Gonyosoma* and *Oreocryptophis*. Another species formerly and erroneously aligned with *Elaphe* was *Oocatochus rufodorsata*, sometimes called the frog-eating ratsnake, red-backed ratsnake, or Chinese garter snake, which shares no natural history identifiers with other ratsnakes, gives live birth, is highly aquatic, and eats fish and frogs as its primary diet. This species is, therefore, not included in this book either.

本书所涵盖的属包括锦蛇属（*Elaphe*）、方花蛇属（*Archelaphe*）、颌腔蛇属（*Coelognathus*）、玉斑蛇属（*Euprepiophis*）、树栖锦蛇属（*Gonyosoma*）和紫灰蛇属（*Oreocryptophis*）。另一个曾被错误归入锦蛇属的物种是红纹滞卵蛇（*Oocatochus rufodorsata*），在中国大陆有时也被称为红点锦蛇。该蛇与其它锦蛇在自然史方面没有共同点，例如其繁殖方式为卵胎生，同时高度水栖，主食鱼类和蛙类。因此，该蛇种本书也将不做涉及。

Scientific names seem to change almost daily as our knowledge of genetic interrelationships is refined, making inconsistencies among names used in various literature frequent. Although new scientific names have been proposed for some of the species found in this book, I have chosen to use the older, more familiar scientific name with the proposed genus name in parentheses. Please refer to the Taxonomy and nomenclature section in the species account for more information on a particular taxon's scientific names.

随着人们对基因相互关系的了解不断深入，物种的学名似乎每天都在发生变化，这使得同一物种的学名在不同文献中经常出现不一致的情况。尽管本书中的某些物种已有新的拟用学名，作者仍选择使用更为人熟知的旧学名，但会将其拟用属名标注在方括号中。欲知某物种学名的更多信息，请参阅该物种介绍章节中的“分类与命名”版块。

A plea for snake conservation in the 21st century / 21世纪的蛇类保护

Reptiles and amphibians are among the most vulnerable vertebrates in the world today. Compared to mammals and birds, which tend to be more appealing to the general public due to their charismatic nature, beautiful plumage or songs, reptiles and amphibians are at a disadvantage. Within this taxonomic group, snakes face many more prejudices than the other families. Snakes are often feared and hated around the world and thus persecuted simply for being snakes. Some species are venomous and kill several people per year worldwide, which has led to a stigma that all snakes are potentially deadly. Many are killed or severely injured when encountered by humans, and entire populations have been systematically eradicated over periods of time, especially those sharing communal hibernacula, such as some populations of rattlesnakes in North America for example.

两栖和爬行动物是当今世界上脊椎动物中最脆弱的群体之一。哺乳动物或鸟类，通过自身的魅力，例如悦耳的歌声或绚丽的羽毛，无疑更能吸引公众的关注。在此方面，两栖和爬行动物完全处于劣势。而在两栖和爬行动物大家庭中，蛇类相比其它成员无疑面临着更多的偏见。蛇类在世界各地被人们恐惧和憎恨，并因此遭受了许多无故的迫害。某些有毒蛇种每年造成多人死亡，这使得所有蛇类都背负着“致命杀手”的恶名。许多蛇在与人类相遇时被残害，一段时间内的种群数量惨遭灭顶之灾，尤其是有群体冬眠习性的蛇种，如北美的一些响尾蛇。

In reality, only a small percentage of snake species, about 7%, are potentially dangerous to humans, but it is challenging to communicate this fact with the general public. Snake populations have been decreasing rapidly across the globe (Zipkin *et al.*, 2020). Snakes suffer from a bad reputation and thus are persecuted by humans out of fear. The most significant factor to population decline comes from loss of habitat. To compound matters, over-exploitation from humans is yet another form of loss, where snakes are sought after for food or traditional medicines, or from individuals being removed from the breeding population to be sold in the pet trade.

实际上，只有一小部分蛇类（约7%）对人类具有潜在威胁，但要宣传普及这一事实却充满挑战。研究表明：全球范围内蛇类的数量正在不断骤减（Zipkin等人，2020）。最主要的原因是栖息地的丧失。然而，人类的过度开发利用，如出于食用、药用、宠物观赏等目的对野生蛇类的捕捉，也令情况雪上加霜。

Human exploitation is significantly damaging in many Asian nations where traditional medicine often calls for various parts of snakes, such as the use of the gall bladders from cobras to treat skin diseases, pain, and intestinal hemorrhage.

在亚洲的许多国家，人类对蛇类的过度利用尤其具有毁灭性的影响。这些国家的传统医学相信蛇的不同部位具有不同的疗效，如眼镜蛇的胆囊被认为可以治疗皮肤疾病、缓解疼痛，甚至应对肠道出血。

Among all of these threats, habitat loss is the most significant contributor. You cannot have aquatic snakes without water or woodland snakes without trees. As humans move away from cities and develop natural areas, they bring predators, especially cats, which prey heavily on local wildlife and impact the surrounding environment. Humans introduce chemical pollutants into the ecosystem, some of which are either directly harmful to snakes themselves or indirectly dangerous through poisoning their food. The fragmentation of habitats to create new roads and increased vehicles on said roads has caused declines in many snake populations living adjacent to busy streets. Birds can fly over roads, and, comparatively, mammals can quickly run across lanes as they have a high metabolism, whereas snakes and other ectotherms frequently suffer high mortality rates when confronted with busy roads, as they are slower to cross.

所有这些威胁中，栖息地的丧失为首要因素。水栖蛇类脱不开水，林居蛇类离不开树。当人类从城市转向对自然区域进行开发时，他们也带去了掠食者，尤其是猫。猫会大肆捕食当地的野生动物并对周围的环境产生影响。人类也导致化学污染物进入当地的生态系统。其中一些对蛇类产生直接危害，另一些则通过污染蛇类的食物而产生间接危害。公路的铺设使原本完整的栖息地支离破碎，无数蛇类更是在车轮下丧生。鸟类可以飞越公路，哺乳动物也相对而言可以快速地穿过车道，剩下蛇类和其它变温动物由于缓慢的移动速度成为最大的牺牲品。

If snakes are to remain a part of our fauna, we must initiate additional conservation measures. Unfortunately, we do not know enough about many of our snake species' life requirements to formulate adequate conservation plans. However, certain needs are apparent. One must protect the lands and waterways harboring substantial or imperiled populations from undue human disturbance and pollution. The single best strategy we can employ to help any struggling species is to give it a place to reproduce. This need is best met through the preservation of habitat. Not only does the preservation of habitat give those species the physical space which they need to reproduce, but it provides an umbrella effect for any other species that may have similar habitat requirements, which in turn provides more resources and a healthier ecosystem for all species involved. There should be a trend away from the use of dangerous residual pesticides and herbicides. The pet trade should be strictly monitored and regulated, and local governments should pass and enforce legislation controlling the capture of snakes in the wild. Responsible collection of wild specimens for captive care, if done correctly, can help aid in snake conservation, but too frequently it is done in an irresponsible and therefore unsustainable way.

如果要将蛇类保留在动物的大家庭中，我们必须采取额外的保护措施。不幸的是，我们对许多蛇类的生存需求了解不足，因而无法为它们制定合适的保护计划。尽管如此，某些需求还是显而易见的。它们生存的水陆环境应免于人类的过度干扰和污染。对任何处境艰难的物种来说，为其繁衍提供合适的空间无疑是最佳的策略。通过保护栖息地无疑可以最大程度地满足这一需求。保护栖息地不仅为这些物种的繁衍提供了所需的物理空间，同一环境下的其它物种也因此受益。如此一来，整个生态系统也变得更为健康。同时，应减少农药和除草剂的使用。宠物交易也应受到严格的监管，地方政府应通过立法来管控对野生蛇类的捕捉。负责任地采集野生标本用于人工饲养，若操作得当，将对蛇类的保护产生益处。然而大多数情况下，此类采集是不负责任且无度的。

Most importantly, people must be educated about the natural history, economic importance, and the plight of snakes. A fostering of awareness in people should make them more interested in protecting snakes and wildlife in general. The creation of a conservation mode toward snakes is a significant purpose of this book.

最重要的是，公众必须对蛇类的自然史、经济意义以及其所面临的困境有所了解。培养人们的相关意识将使他们对保护蛇类乃至其它野生生物更感兴趣。为蛇类构建一种保护模式是本书一大重要写作目的。

Nowadays, there are many social media and community science programs available for an average citizen to help contribute towards conservation. One of the better citizen science programs/ apps for reptiles and amphibians, in my opinion, is HerpMapper (HM; www.herpmapper.org). HM is intended to act as a photographic museum depository. Unlike many other citizen science apps, data sent to HM is protected from the general public and is only made available to conservation agencies upon request and justification. HM is a nonprofit organization designed to gather and share information about reptile and amphibian observations across the planet.

如今，普通市民可通过众多的社交媒体和社区科学项目来为保育工作做出贡献。在我看来，HerpMapper（www.herpmapper.org）是这方面的佼佼者。HerpMapper是一个非营利性组织，旨在收集与共享全球范围内两栖和爬行动物的野外观测记录，功能类似于一个摄影博物馆。与其它面向公众开放的科学应用程序不同，用户向HerpMapper提交的数据将得到保护，保育机构需提交申请并出示相关证明后才可获取信息。

HerpMapper

www.HerpMapper.org

HerpMapper (HM) is a cooperative project, designed to gather and share information about reptile and amphibian observations across the planet. Using HM, you can create records of your herp observations and keep them all in one place. In turn, your data is made available to HM Partners – groups who use your recorded observations for research, conservation, and preservation purposes. Your observations can make valuable contributions on behalf of amphibians and reptiles.

HerpMapper Wants You!

- Data contributors
  - Community scientists
  - Herpers
  - Professionals
- Data users
  - National, regional, and local units of government
  - Non-governmental conservation groups
  - Existing herp atlas projects
  - Researchers
  - Other conservation partners

Share with Confidence!

- Public does not have access to detailed location information
  - County-level only in United States
  - Similar scale for other countries that do not use counties
- Maintain access to data you have submitted
- Export your data in a wide variety of formats (more upon specific request)

Data is Available!

- Free access for research, conservation, and preservation purposes
  - Supports one-time requests and continuous access
- Shared data available at multiple scales
  - City to international-level access
  - Species-specific access
- Photo or audio vouchers for all observations

Easy to Use!

- Easy web interface that extracts voucher metadata when present
- FREE mobile app that allows multiple configurations
- For more information about contributing or receiving data, or for questions and concerns, contact: info@herpmapper.org

HerpMapper

官网地址：www.HerpMapper.org

HerpMapper（简写HM）是一个旨在收集与分享全球两栖和爬行动物野外观测记录的合作性网站。通过使用HM，用户可以一站式保存自己所有的观测记录，而与HM合作的科研机构则可以共享使用这些数据，以便更好地开展两栖和爬行动物研究与野外种群保育工作。

HerpMapper 需要您的加入！

HM的数据提供者包括：

- 民间学者
- 两栖和爬行动物爱好者
- 专职科研人员

HM的数据使用方包括：

- 各级政府机构
- 非政府保育组织
- 两栖和爬行动物科研机构及研究者
- 其他合作的保育机构

在HM您可以分享您的数据！

- 公众无权获悉您的具体观测地点（对公众开放的信息仅细化至“县”层面）
- 随时访问并使用自己所上传的数据
- 可以以多种格式导出已上传数据

真实可用的数据！

- 科研机构、保育团体可免费使用本网站所有数据
- 方便共享
- 支持国际范围内数据分享
- 可直接通过物种搜索分享数据
- 支持图片和音频数据上传

简单的操作！

- 便捷的网页界面方便元数据提取
- 免费手机app支持多平台操作

欲知更多详情，欢迎通过电子邮件联系：info@herpmapper.org

（本网站目前仅支持英文，但后期将推出中文服务）

Captive care: background and history

Among herpetoculturists, the allure of the Asian ratsnakes continually grows due to their beautiful colors and patterns, relative hardiness in captivity, and generally docile nature. Responsible breeding and propagation of species in captivity can help alleviate the pressure put on wild populations. These efforts tend to fail when a species is difficult to keep alive in captivity yet still is desired to be maintained. This results in many failed attempts, specimens dying, and additional specimens being collected from the wild to replace them until success is finally achieved. Depending on the species, this approach is not sustainable. Therefore, it should be understood by all responsible pet owners not to attempt to keep a species beyond one's expertise level. It is also the goal of responsible pet owners to disseminate successful husbandry techniques to stabilize a captive population so that wild animals no longer need to be collected. Since Asian ratsnakes were introduced into herpetoculture, the care of the species in captivity is a subject that has been under-served by many facilities and keepers worldwide. Some facilities and private collectors have failed to successfully breed or even keep some species alive, while others have succeeded. It has only been within the past 20-25 years that herpetoculturists have understood the captive care requirements of these species. Much of this understanding was accomplished through trial and error to successfully determine what environmental conditions were necessary to maintain and propagate this taxonomic group in captivity. A big part of this understanding came through a working knowledge of parasitology and subsequent treatment of these parasites to establish these species successfully in captivity.

人工饲养：背景和历史

由于绚丽的色彩和斑纹，普遍较温顺的性格，以及对人工饲养较强的适应性，亚洲锦蛇在爱好者群体中的人气一路攀升。负责任的人工饲养有助于减轻野生种群所承受的压力。如某一物种难以在人工环境下存活，若仍执意为之，相关努力往往会适得其反。原个体死亡后，饲主为弥补损耗，会采集更多的野生标本。鉴于此，人工饲养也并非一个完全可持续的保护方案。所以，所有负责任的饲主都应该谨记：切勿试图饲养超出自身能力水平的物种！负责任的饲主也应该努力宣传成功的饲养经验。如此一来，人工环境下的种群数量日趋稳定，对野生个体进行采集也就不再必要。自亚洲锦蛇被人工饲养以来，对于如何在人工环境下照料这些物种，世界上许多机构和个人饲养者的做法仍有欠缺之处。一些机构和个人未能成功繁殖甚至养活某些蛇种，但也不乏成功的案例。仅在过去的20～25年间，爱好者才逐步了解这些物种具体的饲养需求。这方面的认知一部分是建立在对理想饲养环境反复试验探索的基础之上，而另一部分则是建立在对寄生虫学的研究以及寄生虫感染的治疗经验之上。

Photographers, photographs, and illustrations

Contributing photographers are credited at the end of the captions that describe the images. If no photographer is cited, then the image is from the author. Unless otherwise noted, all illustrations are by the talented Adeline Robinson (www.adelinerobinsonart.com).

摄影师、照片与插画

本书中部分照片下方注明了照片提供者的姓名；如未作引用说明，则表示照片由作者本人所拍摄。本书中部分插画下方也注明了插画提供者的姓名；如未作引用说明，则表示插画由Adeline Robinson（www.adelinerobinsonart.com）所提供。

Mandarin pronunciation notes for English speakers / readers

This book is obviously not meant to be a language book, but I feel there should be a few basics mentioned, because as I was reviewing literature, I still see several sources using very outdated location spellings which could ultimately lead to confusion. And having a general understanding of some of the more basic pronunciations could help individuals in determining the current locations of some of these places mentioned in literature if one knows the basic rules. Chinese is a tonal language, with the Mandarin dialect having 4 tones. I am not going to discuss the tones and how they are pronounced. For those interested in learning more than the basics below, those individuals could look into some proper language books (which I encourage).

Symbol	Approximate pronunciation	Pronunciation example
c	like ts in pits	cao
ch	like ch in church	chi
h	like the h in honey	hua
q	like ch in chip	qi
r	like r in raw	ren
sh	like sh in wash	shi
x	like sh in she	xin
z	like ds in kids	zi
zh	like j in judge	zhong
a	like a in father	ba
e	like e in her	e
i	1. like e in me,	1. yi
	2. not pronounced after c, s, z	2. ci
	3. after ch, sh, zh, and r, pronounce all letters of the syllable like i in bird	3. chi
o	like aw in awe	wo
u	like oo in spoon	wu
ai	like ai in aisle	ai
an	like an in ran	an
ang	like ang in rang	ang
ao	like ow in how	ao
ei	like ei in eight	lei
en	like en in open	en
eng	like ung in hung	heng
er	like err	er
ia	like ya in yard	xia
ian	similar to ee-an, like Lee-Anne	lian

Symbol	Approximate pronunciation	Pronunciation example
iang	like ee-ang	liang
iao	like ee ee-ow	liao
ie	like ye in yes	lie
in	like the ean in lean	lin
ing	like ing in thing	ling
iong	like ee-ong	xiong
iu	like yo in yoga	liu
ou	like ou in dough	lou
ong	similar to own	long
ua	like wah	hua
uai	similar to why	huai
uan	like wahn	huan
uang	like wahng	huang
ue	like u + eh	xue
ui	similar to way	hui
un	like uan in truant	hun
uo	similar to war	huo

Species accounts layout / 物种介绍布局

Each species account has the following structure / 本书对各物种的介绍将包含以下版块：

Scientific name: One of the main reasons Latin names are preferred in scientific literature is that they are relatively consistent and standardized internationally. Unlike common names, which vary by region or local culture, a scientific name does not vary geographically. Scientific names, however, *may* vary depending on which authority someone follows. Depending on the species involved, some scientific sources cannot agree on the placement of a specific taxon, which results in a species being assigned to one genus by one authority, and another genus by the other source. To overcome this issue, in each account, I list the scientific name while including in [brackets] the proposed name, if appropriate, followed by the investigator who initially described the species in (parenthesis).

An example would be the Beauty ratsnake: *Elaphe [Orthriophis] taeniura* (Cope, 1861). This ubiquitous species' current scientific name is *Elaphe taeniura*, but some researchers suggest the species be placed in the *Orthriophis* genus. The scientist who initially described the species was Edward Drinker Cope, and the year he did so, in 1861.

学名：拉丁名在科学文献中更受青睐的主要原因之一是其在国际上相对一致且标准化。同一物种的常用名在不同地域或文化间可能截然不同，但学名却不会随着地域的变化而有所改变。然而，同一物种的学名却可能因为遵循不同的权威而有所不同。如果学界未就某一具体分类达成一致，这就会导致某物种被不同的权威归隶不同的属。为解决这一问题，本书对各物种的介绍都将包括学名，适当情况下在方括号“[]”中列出拟用属名，之后在圆括号“()”中列出首次描述该物种的研究者。以“*Elaphe* [*Orthriophis*] *taeniura* (Cope, 1861)”为例，该物种目前的学名为*Elaphe taeniura*，但已有专家建议将其改隶晨蛇属（*Orthriophis*），首次对该物种进行描述的人士为Edward Drinker Cope，时间为1861年。

English name: Unlike scientific names, English names are not standardized and vary across the world. In this section, I provide examples of some English names, if available. As mentioned previously, people in different regions tend to have their local names for the species in question. For example, the English language name for *Oreocryptophis porphyraceus* is the Bamboo ratsnake. It is called the Red mountain racer in Hong Kong; in Taiwan, it is called the Red bamboo ratsnake; it is called the Banded trinket snake in Vietnam and India. Taxonomically, at the species level, they are all the same snake.

英文名：与学名不同，同一物种的英文名在全球范围内并不统一。本版块将列出某物种的数个英文名。但正如上文所提及的，同一物种的英文名会因地域的不同而有所不同。例如，紫灰锦蛇（*Oreocryptophis porphyraceus*）的英文名为“Bamboo ratsnake”。在中国香港，该物种的英文名为“Red mountain racer”，在中国台湾则为“Red bamboo ratsnake”，在越南和印度又作“Banded trinket snake”。从分类学上来看，它们全部为同一物种。

Chinese name: Like the use of English names, the Chinese names may also vary with a species' location. For example, *Oreocryptophis porphyraceus* has three different Chinese names, depending on whether it is found on the mainland (紫灰锦蛇), in Hong Kong (紫灰山隐蛇), or Taiwan (紅竹蛇). This book will only use the common name for a species for convenience, as used in mainland China. It would be cumbersome to include all three names as "official," especially considering that Taiwan and Hong Kong use Traditional Chinese writing instead of Simplified Chinese, which is used on the mainland. For those names that differ from the mainland, they will be provided in the **Comments** section of each species account in Traditional Chinese. As taxonomy is still constantly changing in Asia, many species are being reassigned at the genus level, and following Chinese nomenclature rules, Chinese names need to be closely associated with the genus the species in question belongs to. If a species has been placed into another genus, its Chinese name will also be changed accordingly. Given the constant changes in the taxonomy of China's herpetofauna, the same species often has different Chinese names in varying literature. The Chinese names in this book follow those in the Updated Checklists of Amphibians and Reptiles of China (Wang *et al.* 2020).

中文名：情况和英文名类似，不同地域间的称呼也不尽相同。例如，*Oreocryptophis porphyraceus* 在中国大陆被称为“紫灰锦蛇”，在中国香港则被称为“紫灰山隱蛇”，在中国台湾又作“紅竹蛇”。考虑到中国香港和中国台湾仍使用繁体中文，简单起见，本书仅采用中国大陆方面对某蛇种的称呼作为中文名。若某物种在中国香港和中国台湾的用名与中国大陆有所不同，那港、台地区的用名将以繁体中文标注在各物种介绍章节中的“注释”版块。此外，亚洲许多物种的具体分类仍充满变数，某一物种的中文名按照命名规则又常与其所归隶的属有关。若该物种改隶它属，其中文名也会体现相应的变化。由于中国两栖和爬行动物的分类体系仍在不断修订和完善之中，同一物种的中文名在不同文献中常出现不一致的情况。本书中物种的中文名遵循王剀等人于2020年发布的《中国两栖、爬行动物更新名录》。

Etymology: Etymology is the study of the origin and evolution of words. Scientific names are composed of Greek and Latin root words. When a species is described, most authors will include an etymology section describing why that animal was given that name. Often it is a description of color, or pattern, or some aspect of the natural history of that species. An example is *Boiga dendrophilia: dendro* is Greek for "tree", and *philia* means "loving." When the two are combined, "tree-loving" refers to this species' natural history, a highly arboreal species. Other options include naming the species after a geographical location or in honor of someone contributing significantly to the field. This is called a patronym. An example is *Thermophis zhaoermii*: Zhao Ermi (1930 – 2016) is considered the father of Chinese herpetology. This species was named in his honor due to his contribution to the field of Chinese herpetology.

词源：顾名思义，本版块是对词起源和演变的研究。学名由希腊语和拉丁语词根所组成。当描述某一物种时，大部分作者都会在词源部分解释某物种缘何被赋予该特定名称。通常，学名是对某物种颜色、斑纹或自然史某些方面的描述。以“黄环林蛇（*Boiga dendrophilia*）”为例。“*dendro*”在希腊语中意为“树”，“*philia*”意为“喜欢…的”，两者相结合，意为“喜欢树的”，该物种高度的树栖性从名字中就可见一斑。有时也可以以某一地点或是以对某领域做出重大贡献的某人的名字来命名。这方面可以以“四川温泉蛇（*Thermophis zhaoermii*）”为例。赵尔宓院士（1930–2016)被认为是中国两栖和爬行动物学之父，因为他对学科发展所做出的杰出贡献，该物种以他的名字来命名。

Terra typica: *Terra typica* is Latin for "Type Locality." In this section, I provide the type locality of the holotype (also known as the type specimen) if it is known. Some authors offer more precise locality data than others. For example, in 1884, when Sauvage described the species *Elaphe davidi* based on specimens collected by Pere Armand David, he listed the type locality as "China." Today, such a locality descriptor is not helpful. The type locality is of interest to scientists because if they want to compare other specimens of the same species, they must do so using the type specimen as a reference point. The type specimen is the original description, and any differences should be made from that type specimen. Having a detailed type locality provides researchers with additional material to compare specimens as sometimes the holotype may be lost to science.

模式产地："*Terra typica*"，拉丁语，意为"模式产地（Type Locality）"。模式标本的采集地点为模式产地。模式产地的精确度取决于物种描述者。例如，Sauvage于1884年描述由Pere Armand David所采集的团花锦蛇（*Elaphe davidi*）标本时，Sauvage所标注的模式产地为"中国"。很显然，时至今日，如此精确度已没有太大的参考价值。科学家都对"模式产地"这一信息极感兴趣。一旦他们想比较同一物种的其它标本时，则必须以模式标本作为参考。因为模式标本代表最初始的描述，所有差异都应该建立在与模式标本比较后方可得出。在模式标本缺失的情况下，精确的模式产地有助于学者再次找到其它的标本从而进行相关的比较。

Length: Where information is available, the size at sexual maturity, the size at hatching, and the maximum reported size will be reported. If more detailed information is available, such as differences between the sexes in sizes, those are reported as well. All sizes listed are in total length (TL) unless otherwise noted. If available, tail lengths are also provided, following the total length.

体长：在相关信息可获取的情况下，本版块将列出某蛇种性成熟时的尺寸，刚孵化出壳时的尺寸以及已知的最大尺寸。若可获取更详细的信息，还会列出雌雄两性间的尺寸差异。如未作特殊说明，本书中的"体长"均指身体全长。若可能，本书还将提供尾长数据。

Description: This section will cover the physical description of the species, including diagnostic characteristics, such as scale counts, pattern, coloration, head and body shapes, pupil characteristics, and other biological information that describes and distinguishes the species from other members of the genus.

外观描述：本版块旨在涵盖对某物种的各类外观描述，包括辨识特征、鳞片计数、斑纹、颜色、头部及身体形状、瞳孔特征，以及其它有助于在外观上描述及辨识某物种的生物信息。

Distribution: In this section, a list of provinces where the species is found within Greater China is provided, and for those occurrences outside of Greater China, I give a list of countries. A distribution map is also provided with as much detail as possible based on what can be found in the literature and my observations. Depending on data resolution, some distribution data are at the provincial level, especially in places infrequently surveyed. Where possible, and where available, I use data at county or city level specificity as well. It should be recognized that there are many disputed boundaries, and regardless of what the international assessment is, since this is a book based on "Greater China," it seemed important to include any boundaries that are in dispute. No one can predict any boundary agreements that could be made in the future and this way the book can be as inclusive as possible. I hope the boundaries illustrated in this text do not offend anyone, and if so, I apologize. If a species is present in both Hong Kong and Taiwan, I provide insets to both of those regions to give more clarity to the distribution. If neither inset is displayed, then it means that species has not been documented in those areas.

地理分布：本版块将列出某物种在中国所分布的省份以及在中国以外有所分布的国家。基于文献信息及个人观测，本书还将提供分布图并尽可能做到详实。根据数据颗粒度，某些分布数据仅细化至"省"，尤其是在调研较少的地区。其它情况下，颗粒度将尽可能细化至"县"或"市"的层面。毋庸置疑，部分领土或有边界争议，无论国际社会对此作何评定，考虑到这是一本基于中国的书籍，故似乎有必要将任何有争议的领土都包括在内，毕竟无人能预测各方未来还会在此方面达成何种协议。如此一来，本书在物种的覆盖上也能做到最大程度的兼收并蓄。若书中的地图造成任何误会，作者本人在此深表歉意，还望各位海涵。如果某物种在中国香港和中国台湾皆有分布，作者还将为这两个地区提供单独的分布图，以使分布信息更为清晰。反之则说明某物种在上述两地未有分布记录。

Habits and habitats: This section covers other aspects of the ratsnakes' life histories, such as activity, habits, habitats, and how they respond to threats.

习性与栖息地：本版块旨在涵盖某物种生活史的其它方面，例如活动规律、习性、栖息地以及如何应对威胁。

Diet and Feeding: This section will describe what the ratsnake species eat, preferably in its natural habitat. If in vivo observations or postmortem evidence from wild specimens are lacking, information about what the species is known to eat in captivity is provided. Still, in that case, the information does not necessarily represent that animal's natural food source. Further research to answer that question is encouraged.

食性与喂养：本版块旨在介绍某物种以何为食，重点关注其在自然环境中的食性。若野生环境下的食性不详，本书将提供某物种在人工饲养下的食性信息，但人工饲养下的食性不等同于野生环境下的食性，更专业的解答仍待未来进一步的研究。

Captive care: This section describes both general and specific requirements regarding the captive husbandry of each species. This information comes from the current literature. If the information is not published, it originates from personal communications with individuals or zoos that have successfully cared for the species.

人工饲养：本版块将介绍某物种在人工饲养时的基本及特殊需求。相关内容源自于当下的各类文献。若此类文献尚未发布，相关信息则来自于作者与成功饲养该物种的个人或机构的私下交流。

Reproduction: This section provides information on reproduction, such as breeding season, clutch sizes, incubation periods, and temperatures. Data in this section may come from both captive and natural observations.

繁殖：本版块旨在提供繁殖方面的有关信息，例如繁殖季节、窝卵数、孵化期和孵化温度。本版块的相关数据源于人工和野生环境下的观测。

Taxonomy and nomenclature: This section is meant to provide any recent taxonomic changes, alternative scientific names, and any controversies about them are discussed in this section, as are any recognized subspecies in the surrounding countries. Distributions, type localities, and descriptions of the subspecies are also provided in this section.

分类与命名：本版块旨在介绍某物种近年来的所有分类变化、其它替代性的学名以及对此的各式争议。此外，若某物种下的某些公认亚种在中国周边国家有所分布，这些亚种也将在本版块进行讨论。亚种的分布、模式产地以及外观描述也将一并在本版块进行介绍。

Conservation status: This section describes the species' conservation classification according to the International Union for Conservation of Nature (IUCN) Red List of Threatened Species. The IUCN was established in 1948 and has since become the world's most comprehensive information source on wildlife species' global extinction risk status. It should be emphasized that the keyword in the previous sentence is "global." The IUCN only assesses species according to their global risk, not their risk in any one country they may occur in. The IUCN Red List uses the following ranks in assessing species:

保护状况：本版块根据世界自然保护联盟《濒危物种红色名录》对物种的保护分级情况进行介绍。世界自然保护联盟创立于1948年，现已成为野生物种全球灭绝风险方面最全面的信息来源。应强调的是，世界自然保护联盟仅根据某物种在全球的灭绝风险来进行评估，而非根据某物种在任一国家的灭绝风险来进行评估。《濒危物种红色名录》在评估物种时使用以下级别：

NE means Not Evaluated. No biologists have been assigned to review the species in question to see if it warrants protection or not.

未予评估 / Not Evaluated（NE）：世界自然保护联盟尚未对某一物种是否需要保护进行评估。

DD means Data Deficient. DD means the species has been examined, but the researchers reviewing the species could not find enough information on the species to assess its threat level accurately, such as not having a clear understanding of a species' population trend.

数据缺乏 / Data Deficient（DD）：已对某物种做了相关研究，但研究发现相关信息较为匮乏（如对某物种的种群趋势缺乏清晰的认知），无法准确评估其受威胁程度。

LC means Least Concern. A species in this category usually has a robust population, is considered "common," and therefore does not face any significant global threats to its population.

无危 / Least Concern（LC）：该等级下的物种通常种群数量庞大，未受到任何重大的全球性威胁，被视为"常见"物种。

NT means Near Threatened. A species in this category is at risk of becoming threatened if trends in its population continue in the current trajectory. This usually happens with a species when there has been a consistent reduction in population numbers or habitat over time.

近危 / Near Threatened（NT）：若不采取相关措施，未来一段时间后某物种接近符合或可能符合受威胁等级。通常适用于那些种群数量持续减少、栖息地不断缩水的物种。

VU means Vulnerable. A species in this category has a high risk of becoming endangered in the wild.

易危 / Vulnerable（VU）：尚未达到"濒危（EN）"等级，但某物种野生种群面临濒危的机率较高。

EN means Endangered. A species in this category has a high risk of becoming extinct in the wild.

濒危 / Endangered（EN）：尚未达到"极危（CR）"等级，但某物种野生种群面临灭绝的机率较高。

CR means Critically Endangered. This category has an extremely high risk of extinction in the wild and is represented by a small population. To be classified in this category, according to the IUCN, a species must fit ANY of the following criteria: "1) Population size reduction, usually by at least 80%. 2) Reduction across a geographic range: This reduction must occur over less than 100 km2, or the area of occupancy is less than 10 km. 3) Population decline: The population must decline to less than 250 mature individuals and either have a decline of 25% over three generations or ten years, or have extreme fluctuations, or over 90% of mature individuals in a single subpopulation, or no more than 50 mature individuals in any one subpopulation. 4) Population size reduction: The population size must be reduced to numbers of less than 50 mature individuals. Lastly, 5) Probability of extinction: There must be at least a 50% probability of going extinct in the wild within three generations or ten years, whichever comes first."

极危 / Critically Endangered（CR）：该等级下物种的野生种群面临灭绝的几率极高，数量已岌岌可危。归入该等级的物种需满足以下五大标准中的任一条：①.种群数量至少减少80%。②.分布区小于100平方公里或占有面积小于10平方公里。③.种群成熟个体数量小于250且符合以下四点中任一条：a.预计三个世代内或10年内，成熟个体数量将持续至少减少25%；b.成熟个体数量极度波动；c. 90%的成熟个体集中于某一亚种群中；d.某一亚种群成熟个体数量不足50。④.种群的成熟个体数量低于50。⑤.今后10年内或三个世代内，野外灭绝的几率至少达到50%。

EW means Extinct in the Wild. A species in this category can still be found in captivity, but it is no longer found in its natural habitat and known geographical range. One example would be the Père David's Deer (*Elaphurus davidianus*).

野外灭绝 / Extinct in the Wild（EW）：某物种仅生活在人工圈养、栽培条件下，其自然栖息地已无种群分布，例如麋鹿(*Elaphurus davidianus* / 俗名"四不像"，原产于中国长江中下游沼泽地带)。

EX means Extinct. There are no living specimens known.

灭绝 / Extinct（EX）：某物种在地球上已无已知存活个体。

Similar species: Not all entries will have this subheading. Some snakes of different genera and species look remarkably alike. In this section, I try to provide some examples of species that could be mistaken for the target species and provide enough information to distinguish the target species from those with a similar appearance.

相似物种： 由于某些来自于不同属的不同蛇种在外观上极为相似，本版块将列举出与所介绍的物种极易混淆的相似种，并对如何区分做出说明。本书中部分物种的介绍章节将包含本版块。

Comments: Not all entries will have this subheading. This is the last section under each species account and is reserved for any miscellaneous information that was not included in any previous categories. If there is no miscellaneous information to mention, this subheading will not show up.

注释： 所有未归入先前版块的其它信息将包含在本版块。本书中部分物种的介绍章节将包含本版块。

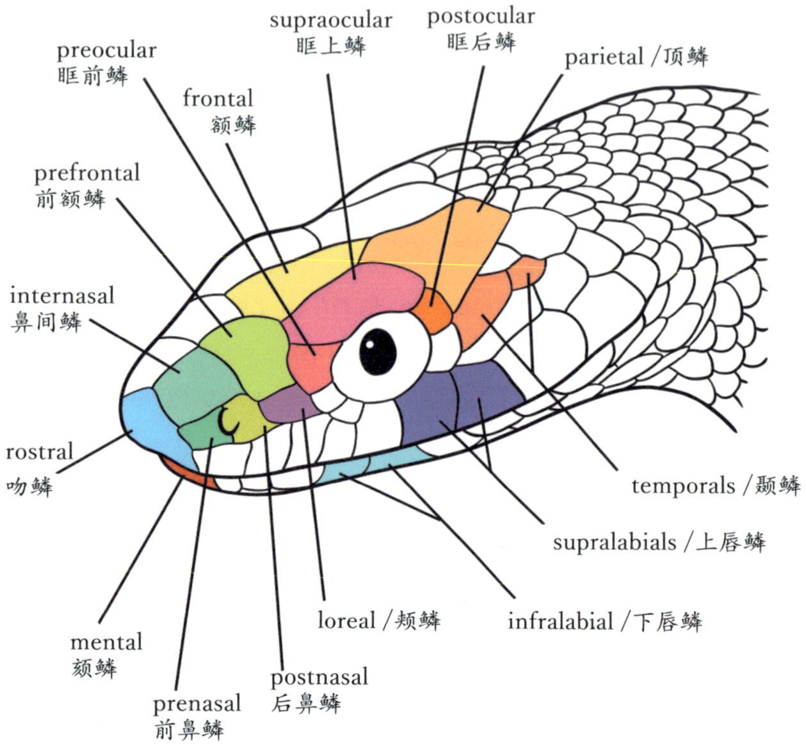

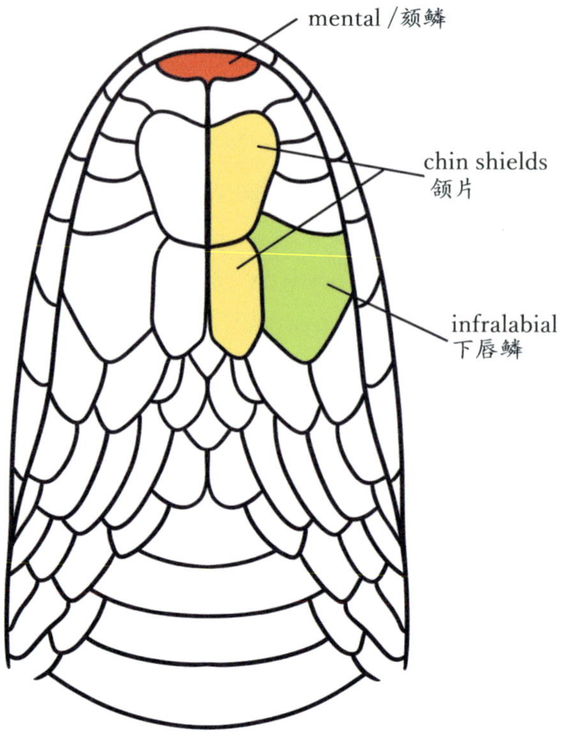

Scalation: Identifying and counting the number and arrangement of cephalic scales, as well as the condition (single or paired) of the anal scale and subcaudals scales is among the easier ways to help distinguish one species of ratsnake from another. When these data are coupled with dorsal scale counts, it further helps to distinguish species from species. One should not rely on just pattern and coloration because many species can have mutations that may be drastically different from the standard color or pattern. Though scalation may also be affected by mutations, it is less common comparatively.

鳞被：对头部鳞片、肛鳞及尾下鳞的数目进行计数和对其排列方式和外观进行鉴识是区分各锦蛇物种的捷径之一。以上数据与背鳞计数相结合后更能提供进一步的分辨依据。由于许多蛇种存在体色或斑纹变异的情况，故基于体色和斑纹进行判断或导致误鉴的发生。尽管鳞被特征也会受变异情况的影响，但相对而言较不常见。

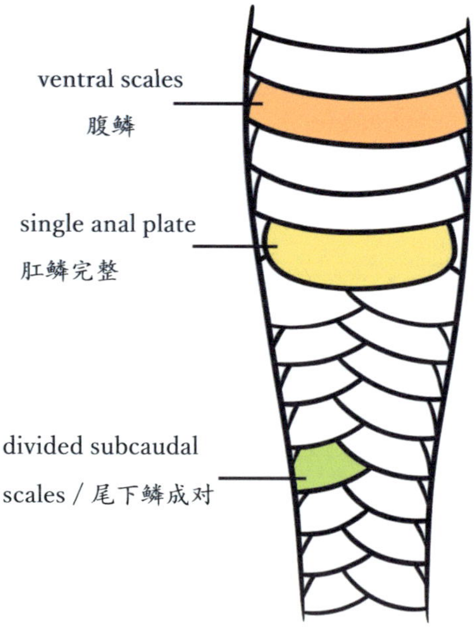

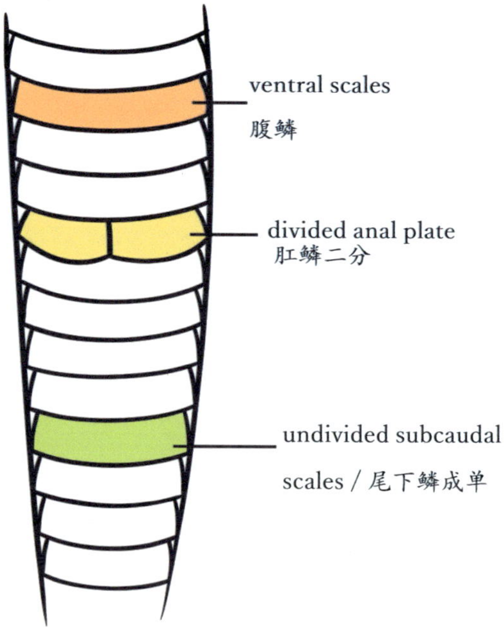

Determining the number of dorsal scales is one of the primary methods for species determination. The illustration below shows three different strategies for counting dorsal scale rows. Typically, when describing a snake, dorsal scale rows are counted anteriorly (just posterior to the neck region), midbody (the middle of the body), and posteriorly (just anterior to the anal plate). One example of a scale row count would be reported as 17:19:19. Scale rows are almost always odd numbered.

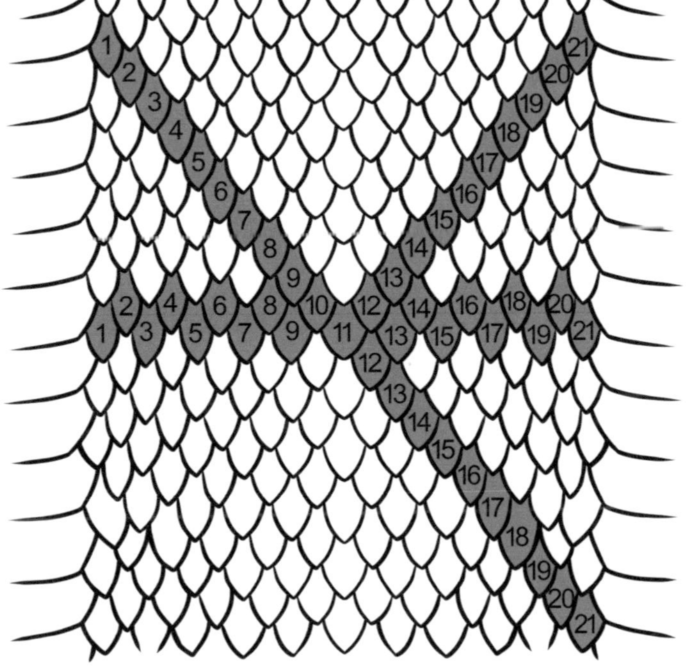

Smooth scales. Note the even surface and absence of ridges or texture.

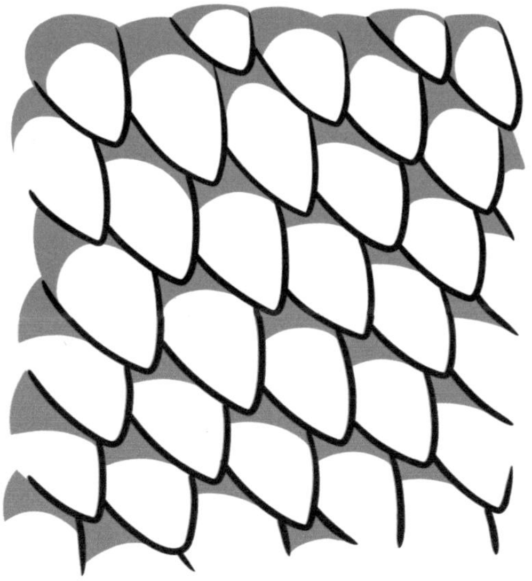

Keeled scales. Note the ridge of the keel in the center of the scales.

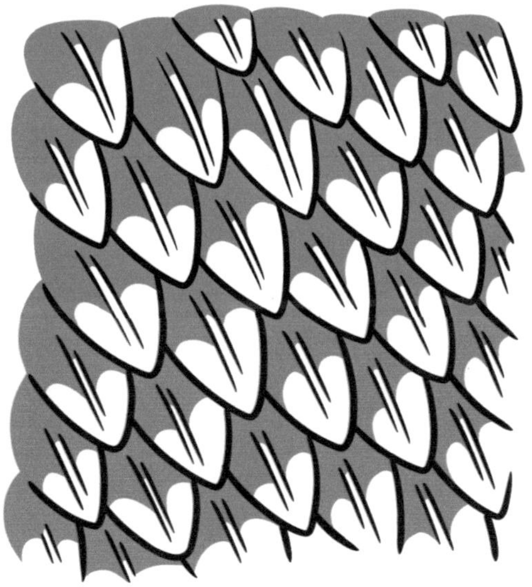

对背鳞进行计数是判定某蛇种的主要手段之一。最上方的示图介绍了三种不同的背鳞行数计数方法。通常而言，计数一般取体前（颈后区域），中段（身体中间）及体后（肛鳞前）3个数据。可以以式表示，如写作“17：19：19（本书中文部分写作：17-19-19）”。背鳞行数一般都是奇数。

Scales on ratsnakes come in three forms. Smooth scales, which lack any type of keel. Feebly keeled, in which the dorsal most scales usually have a slight keel, and the scales become increasingly smooth as you move laterally away from the midline. The final variety is rough or keeled scales, in which each scale, including those on the lateral sides, have a strong, prominent keel. Snakes that are smooth scaled often have a glossy or shiny appearance whereas snakes that are strongly scaled are usually matte in appearance. Examples of the differences in appearance between rough or keeled scaled species to smooth scaled species. All species in this example are North American species and were selected for illustrative purposes only.

锦蛇的鳞片有三种形态：平滑（无棱）、微弱起棱（大部分背鳞起微棱，正中起棱最为明显，随着向体两侧的延展鳞片趋于平滑）和强烈起棱（即使体两侧的鳞片亦起棱明显）。背鳞平滑的蛇种外观通常光滑亮泽，背鳞强烈起棱的蛇种外观则较粗糙无光泽。鳞片不同起棱程度的对比如本页图片所示。图中皆为北美蛇种，在此仅作说明用途。

Apical pits on an albino *Elaphe anomala* (Matthew Most) in order to show the pits more easily (above). Often times in naturally colored species, the apical pits are more difficult to see (below). The more colorful the pattern, usually the more difficult to see the apical pits.

上图为白化赤峰锦蛇（*Elaphe anomala*）的背鳞，注意其上的端窝。相较于正常个体（下图），白化个体背鳞上的端窝更为明显。一般来说，某蛇体背斑纹的颜色越丰富，端窝也越不明显。

Glossary and Abbreviations

Many scientific terms and words are composed of Greek and Latin roots, prefixes, and suffixes. To novice scientists, many of these words may be unfamiliar. One of the benefits of Greek and Latin, aside from not changing in time like many languages (same reason scientific names are presented in Greek or Latin), is that many scientific terms can be chained together to create new words. For example, **dorsal** pertains to the back. A dorsal view looks at the animal from the back. **Lateral** pertains to the side. A lateral view looks at the animal from the side. The two words can be combined as in **dorsolateral** to mean a view that is halfway between a side view and a top-down view. The following is a list of some of the more commonly encountered terminology in herpetology. As this book focuses on China, I also include some definitions of Chinese terminology.

术语与缩写

许多科学术语和单词都是由希腊语或拉丁语词根、前缀或后缀所组成。初级科学家或对其中的部分词略感陌生。除了不会随着时间的推移有所变化外，希腊语和拉丁语的另一大好处是许多科学术语之间可以重新组合以构造新的单词。例如，“dorsal”意为“背部的”。“a dorsal view”意为“背视（从背面角度进行观测）”。“lateral”意为“侧面的”。“a lateral view”意为“侧视（从侧面角度进行观测）”。“dorsolateral”一词就由以上两个单词所构成。“a dorsolateral view”顾名思义表示观测角度介于背视和侧视之间。以下列出两栖和爬行动物学中一些最常用的术语。由于本书重点关注中国，故一些中文术语的定义也包含在内。

A-; An-: Lat. negative prefix that means “not” or something similar that may reverse the action or meaning of the word they attach to / 拉丁语源否定前缀，表示“非，无，缺乏”等否定概念。在单词前加此前缀常构成与该词意义相反的新词。

Adult: When an individual is sexually mature / 成体。个体已达到性成熟.

Aff :Lat. abbreviation for *affinis*, meaning “closely related to” or “akin to” / “affinis”一词的拉丁缩写，意为“近缘的”。

Allopatric: Occurring in separate locations; in opposition to **Sympatric** / 在不同地区生存的。“Sympatric”的反义词。

Anal scale/ plate: Terminal ventral scale, immediately anterior to the cloaca and tail / 肛鳞。最末1枚腹鳞之后，紧覆于泄殖腔上方的鳞片。

Anterior: Near the front; in the direction of the head; in opposition to **Posterior** /身体前部的（靠近头部）。“Posterior”的反义词。

Apical pit (scale pit): In dorsal scales, mostly paired terminal pores with likely tactile function. Their presence or absence is an important systematic character / 端窝。位于背鳞之上，通常为成对的小窝，或具触觉功能，有分类鉴别的参考意义。

Aquatic: Lives in the water / 水生的，水栖的。

Arboreal: Lives above the ground, usually in trees or high shrubs / 栖息于地面之上的（通常生活在树上或较高的灌木丛中）。

Bi-: Lat. prefix for two, twice, or a pair of / 拉丁语源前缀，意为“两个，双，对，两次”。

Brumation: A prolonged period of low metabolic state, but not as deep of a state as hibernation. When animals are brumating, they are still active enough to drink and opportunistically bask / 冬化。长时间处于低代谢状态，但深度不及冬眠。当动物处于冬化状态时，它们仍有足够的活跃度进行饮水或晒太阳等行为。

Cathemeral: Species with no well-defined activity pattern. The species could be found diurnally, nocturnally, or be crepuscular. Being active at sporadic or irregular intervals / 形容某物种无明确的活动规律。其可以在白昼、夜间或晨昏进行活动。

Cephalic: Relating to the head or head end of the body. Situated on, in, or near the head. Cephalic is synonymous with cranial. "Cephalic" comes from the Middle French "cephalique," which was derived from the Latin "*cephalicus,*" which was derived from the Greek "*kephaloikos*" meaning "head" / 头部的。"cephalic" 源于中古法语中的 "cephalique" 一词。再往前追溯，"cephalique" 源于拉丁语中的 "*cephalicus*" 一词。"*cephalicus*" 又源于希腊语中的 "*kephaloikos*" 一词（意为头部）。

cf : Latin abbreviation for "confer" (compare): Used in nomenclature if the specific affiliation is probable. Such as, *Sphenomorphus* cf *indicus,* means a form supposedly belonging to *S. indicus* / "confer（比较）" 一词的拉丁语缩写，常用于物种的命名，指可能存在特定的从属关系。例如，"*Sphenomorphus* cf *indicus*" 指某物种与 "*Sphenomorphus indicus*（印度蜓蜥）" 十分相似，可能为同一物种但并不确定，具体情况仍待进一步的考证。

CIB: Chengdu Institute of Biology, located in Chengdu, Sichuan Province / 中国科学院成都生物研究所。

CITES: Convention on International Trade in Endangered Species of Wild Fauna and Flora / 《濒危野生动植物种国际贸易公约》。

Chang Jiang: Also known as the Yangtze River by foreigners and many maps / 长江。国外地图也经常标注其为 "Yangtze River"。

Chrom-: Gr. *khromatikos* = relating to color / 前缀，源于希腊语中的 "*khromatikos*" 一词，意为 "有关颜色的"。

Cloaca: A common opening where both excretory and genital products exit the body / 泄殖腔。兼具生殖和排泄的功能。

Clutch: Total number of eggs laid by a female at a time / 窝卵数。指雌性一次产卵的数量。

Communal: Lives in groups / 群居的。

Congener: Animal of the same genus / 同属的物种。

Conspecific: Animal of the same species / 同种的。

Copulation: Sexual intercourse / 交配，交尾。

Crepuscular: Active at twilight hours, dawn and dusk / 在黎明或黄昏时分活动的。

Dermal: Of the skin / 皮肤的。

Di-: Gr. prefix for two, twice, or pair of / 希腊语源前缀，意为 "两个，一对，两次"。

Diastema: A space or gap between two teeth / 两齿间的间隙。

Diel: Referring to a 24-hour day, in contrast to diurnal which refers to the time between sunrise and sunset / 二十四小时的。

Diurnal: Active during the daytime, in opposition of **Nocturnal** / 白昼活动的，日行性的。“Nocturnal”的反义词。

Dorsal; dorsum; dorso-: Relating to the back of the body; in opposition to **Ventral** / 背部的；背部。“Ventral”的反义词。

Ectotherm: An animal depending predominantly on external heat sources, primarily the sun / 变温动物，一类主要靠从外界环境中吸收热量来提高自身体温的动物，主要热源为太阳。

Elliptical: Oval-shaped, not round / 椭圆的。

Endemic: Restricted to a particular region / 限于某一特定区域的。

Ery-: Lat. for "red." For example, erythrocyte = red blood cell; erythristic = red coloration / 拉丁语源前缀，意为“红色，红色的”。如“erythrocyte（红细胞），erythristic（红色）”。

Fauna: The total of animal life of a certain region or time. For example, herpetofauna = the total of reptiles and amphibians of a certain region or time / （尤指某区域或某时期的）动物群。例如，“herpetofauna”指某区域或某时期的两栖和爬行动物群。

***Fide*:** Lat. for "confidence" or "in good faith" / 拉丁语，意为“真诚（地），善意（地）”。

Fossorial: Adapted for digging and living underground / 适于掘地和居于地下的。

Genus: Assembly of closely related species / （分类学的）属。

Gr. : Abbreviation for Greek, in reference to the origin of words / “Greek（希腊语）”的缩写。

Gular: Pertaining to or located on the throat / 咽喉的，喉部的。

Hatchling: A newly hatched neonate that emerged from an egg. Sometimes abbreviated as Y.O.Y. (Young Of Year). This age class is usually reserved for the first season of life, prior to brumation. By most definitions, during the following spring emergence, after brumation individuals are classified as **juveniles** / 幼体。在本书中，该词所对应的年龄段一般始于个体出壳，止于个体进入第一年的冬化。第一年的冬化结束后，个体步入小亚成体（Juvenile）阶段。

Hemipenis: One of the two male copulatory organs in lizards and snakes. Plural: **Hemipenes** / 半阴茎。蛇和蜥蜴雄性个体的交配器官为一对半阴茎（复数形式：Hemipenes）。

Herpetology: The branch of zoology that studies reptiles and amphibians. From the Gr. *herpein* meaning "to creep" and *herpeton* meaning "reptile" and *-ology*, meaning "the study of." When Carl Linnaeus coined the term, he lumped amphibians in with reptiles, thinking they were closely related. In fact, they are different Classes (Class Amphibia and Class Reptilia), but because of tradition, the field of herpetology has never been broken up into two more appropriate fields of study / 两栖和爬行动物学。研究两栖动物和爬行动物的动物学分支。源于希腊语单词“*herpein*（意为爬行）”和“*herpeton*（意为爬行动物）”；“*-ology*”，后缀，表示“…学科”。当卡尔·林奈（Carl Linnaeus）创造这一术语时，他将两栖动物与爬行动物合为一类，认为两者是非常近缘的。实际上，两者属于不同的纲（两栖纲和爬行纲）。但由于传统原因，两栖和爬行动物学从未细分为两个或以上的研究领域。

Herps: A slang, and abbreviated term for "herptiles," which is the collective term for reptiles and amphibians / “herptiles（两栖和爬行动物）”的缩写。

Hibernation: A period of deep, low, metabolic state, usually reserved for mammals / 冬眠。生物处于低代谢状态，程度超过冬化。一般用于形容哺乳动物。

Holotype: Gr. *holos* = whole and Gr. *typos* = image. A single type specimen upon which the description and name of a new species is based. Also known as the type specimen. The geographic location that the holotype came from is known as the 'type locality' or *terra typica*. See *Terra typica* / 正模标本，亦称模式标本。某新物种据以命名、描述的那一份标本。该份标本的采集地被称为"模式产地"。源于希腊语，"*holos*"意为"完整的"，"*typos*"意为"图像，形象"。

Hybrid: A individual composed of two different species / 两个不同物种杂交的产物。

Hypapophyses: Projections on the ventral side of the spine, used in egg-eating specialists, meant to assist in breaking up ingested eggs / 椎体下突。某些食卵蛇类的脊椎腹面突出，用于帮助压碎吞咽下的蛋。

Hyper-: Gr. for "over" or "above" or "excessive" / 希腊语源前缀，意为"高于…"或"过于…"。

Hypo-: Gr. for "under" or "below" / 希腊语源前缀，意为"在…下，次于，不足"。

Infra-: Lat. root word usually in reference to a lower position, as in "infralabials" / 拉丁语源词根，意为"下，低"，如"infralabials（下唇鳞）"。

Inter-: Lat. root word meaning "between" / 拉丁语源词根，意为"在…之间"。

Inferior: Lat. for "below" / 拉丁语单词，意为"较差的，次要的，低等的"。

Intergrade: An individual composed of different subspecies but the same species / 指同一物种不同亚种间的杂交个体。

Intra-: Lat. root word meaning "within" or "inside" / 拉丁语源词根，意为"在…之内，内部"。

IUCN: International Union for Conservation of Nature. For individual rank definitions, see the "Comments" section on page 2 / 世界自然保护联盟。

Juvenile: Age-class after emerging from first over-wintering brumation period, usually about 4 - 6 months, which in temperate regions, this would normally coincide with springtime emergence in the year following hatching. The age-class is in effect until the class of "subadult," which is the class just prior to being sexually mature / 小亚成体。在本书中，该词所对应的年龄段一般始于个体刚结束第一年的冬化（通常4～6月龄大），止于个体步入亚成体（Subadult）阶段。

Keel: Narrow prominent ridge / 本书中指鳞片上的棱。

Keratinous: Consisting of keratin. Keratin is the structural protein that makes up scales (in reptiles, not fish), feathers, claws or nails, hoofs, horns (in mammals), and hair / 角质的。角质实际主要成分为蛋白质，如动物的羽毛、蹄、犄角、毛发、趾甲等都由角质组成。爬行动物的鳞甲也由角质构成。

Labial: Pertaining to the lips / 唇的。

Lat. : Abbreviation for Latin, in reference to the origin of words / "Latin（拉丁语）"的缩写。

Lateral: Pertaining to the side, in opposite to **Midline** / 侧面的。

Loreal region: Lateral head region between nostril and eye, from Lat. *lorum*, meaning "rein" / 指头侧鼻孔至眼睛之间的区域。源于拉丁语中的"*lorum*"一词，意为"马勒"。

Mental: Referring to the chin (Lat. *mentum*: chin shield) / 颏（下巴）的。本书中指颏鳞。源于拉丁语中的"*mentum*"一词。

Midline: A median line or plane of bilateral symmetry, especially in reference to the body. In opposition of **Lateral** / 使（身体）左右两侧对称的中线或平面。

Monotypic: Gr. *Mono-* = one, Gr. *typos* = image. Containing only one type or species. For example, a monotypic genus, is a genus with one species / 单型的。例："a monotypic（单型的）genus（属）"表示该属下仅有一个物种。"*Mono-*"，希腊语源前缀，意为"一个，单一"；"*typos*"，希腊语单词，意为"图像，形象"。

Musk: A foul-smelling liquid substance produced by the glands near the inside opening of the cloaca, used as a defense against predators, with intentions of making the snake seem unpalatable / 泄殖腔附近的腺体所分泌的恶臭液体，起自卫作用。

Nei Mongol: Synonymous with Inner Mongolia / 内蒙古自治区，简称"内蒙古"。

Nocturnal: Active at night, in opposition of **Diurnal** / 夜间活动的，夜行的。"Diurnal"的反义词。

Nomenclature: Standardized system of denomination (Lat. *nomenclatura*: calling by name, list of names; *nomen*: name; *calare*: call) / 命名法。源于拉丁语中的"*nomenclatura*"一词，"*nomen*"意为"名字"，"*calare*"意为"称呼"。

Nuchal: Pertaining to the back of the neck / 颈背的。

Ocular: Pertaining to the eye / 眼睛的、视觉的。

Ontogenetic: Pertaining to a change or development from juvenile to adult / 个体发育的。同小亚成体阶段至成体阶段的变化有关。

Opaque: The term used when a snake is undergoing the process of shedding its skin. During this period, a snake's eyes will turn cloudy or blue as a milky substance separates the new skin and the old. Following this opaque stage, the snake's coloration and pattern will clear up but still be duller than normal. A few days after this the snake will begin to shed its skin by rubbing against rough objects in order to separate the layers / 蒙眼。形容蛇蜕皮的术语。在"蒙眼"阶段，由于乳状物质将新、旧皮肤分开，蛇的眼睛会变得混浊或发蓝。这个阶段过后，蛇的颜色和斑纹将变清晰，但仍比正常情况下黯淡。几天之后，蛇便会摩擦粗糙物体开始蜕皮。

Orbit: Another name for the eye / 眼睛。

Oviparous: Egg-laying form of reproduction. Oviparous animals lay calcified eggs outside of the body / 卵生的。

Ovoviparous: Live-bearing form of reproduction. Neonates are born still enclosed in their embryonic membranes. See **viviparous** / 卵胎生的。卵胎生蛇类交配后，卵在母体体内发育，之后母体直接产出仔蛇。卵胎生蛇类的卵无坚韧的外壳，但幼蛇出生时仍包裹在胎膜之中。

Pinyin: The Romanization of Chinese characters based on their approximate pronunciation. For each species account, under the Chinese name section, I provide the Chinese text, as well as the pinyin spelling and pronunciation / 汉语拼音。

Post-: Lat. prefix for "after" or "behind" / 拉丁语源前缀，意为"在…之后"。

Posterior: Near back; towards the tail; in opposition to **Anterior** / 身体后部的（靠近尾巴）。"Anterior"的反义词。

Pre-: Lat. prefix word for "before" or "in front of" / 拉丁语源前缀，意为"在…之前"。

Proboscis: An elongate and flexible extension of the snout / 吻端细长而又可弯曲的突出物。

Reticulated: Forming a net-like pattern or reticulum / 网状的。

Rostral: The scale at the tip of the snout; Pertaining to the nose / 喙的。本书中指吻鳞。

Scansorial: Both arboreal and terrestrial / （适于）攀附的。本书中指某蛇种既有地栖性又有树栖性的生活方式。

Scenting: A term in husbandry used for disguising the scent of a prey item with another prey item / 人工饲养中的一个术语，意为以另一种食物的气味来掩盖某种食物本身的气味。

Scute: Horny epidermal shield / 鳞甲。

Semi: Means "half." Example: A semi-aquatic species spends approximately 50% of its time in the water, and 50% of its time out of water / 表示"半"。例如："semi（半）-aquatic（水栖的）"表示某物种的生活方式为半水栖。

Sensu amplo: "in a relaxed, generous (or 'ample') sense. Similar to *sensu lato* / 极广义的，从极广义上讲，类似于"*Sensu lato*"。

Sensu lato: "in the broad sense." Abbreviated as *s.l.* / 广义的，从广义上讲。缩写为"*s.l.*"。

Sensu stricto: "in the strict sense." Abbreviated as *s.s.* or *s.str.* / 狭义的，从狭义上讲。缩写为"*s.s.*"或"*s.str.*"。

Serrated: Having a saw-like or jagged edge / 锯齿状的，有锯齿的。

Shanxi / Shaanxi: This entry is included because to the casual observer, not familiar with Chinese geography, one may think one or the other is a typo. These are two different provinces in north central China. The double 'A' in Shaanxi is meant to distinguish, in pronunciation, between the two provinces / 山西省（Shanxi）和陕西省（Shaanxi）。两者的英语拼写仅一个字母之差，较易互相混淆。

Shields: A large scale in snakes, lizards, and turtles / 蛇、蜥蜴以及龟类身上较大的鳞片。

Species: Basic systematic unit: An assembly of specimens which resemble each other, occupy similar ecological niches, and can interbreed. Closely related species form a genus. The spelling "species" is both singular and plural / 物种。生物分类学研究的基本单元。同一物种的所有成员在形态上极为相似，占据相似的生态位，并可以互相交配繁殖。近缘的物种归合为"属"。值得注意的是，"species"一词在英语中单复数同形。

Speciation: Evolutionary process that leads to the formation of a new species / 物种形成。

Striations: Having lines along the body / 身体具条纹。

Sub-: Lat. root word usually meaning "below," such as in subocular scales, or subcaudal scales / 拉丁语源词根，意为"在…之下"，例如"subocular scales（眶下鳞），subcaudal scales（尾下鳞）"。

Subadult: A very subjective age class. Subadults are still classified as **juveniles**, but they have started showing characteristics of becoming sexually mature. In most situations, this age class would be in effect as the individual is going into its final brumation period before emerging the next spring as a sexually mature **adult** / 亚成体。该词所对应的年龄段有较大的主观性。一般来说，此年龄段的个体已开始表现出某些性成熟的特征。

Subcaudal: Ventral scales below the tail, posterior to the anal plate / 尾下鳞。尾腹面的鳞片，位于肛鳞之后。

Subspecies: Geographic variations of a species which inhabit different, usually contacting, areas and form intergrades at their contact zones / 亚种。亚种是指同种内由于分布地域的不同（这些地域通常有所重叠）而形成的在形态、特征、性状上有明显区别的生物群体。

Supra-: Lat. root word usually in reference to a higher position, such as in supralabials / 拉丁语源词根，意为"在…之上，超越"，例如"supralabials（上唇鳞）"。

Superior: Above / 上级的，优秀的。

SVL: Snout-to-vent length / 吻肛长度。指的是从吻端至肛孔的长度。

Sympatric: Inhabiting the same area; in opposition to **Allopatric** / 在同一地区生存的。"Allopatric"的反义词。

Synonym: In taxonomy, synonyms are different scientific names of the same taxonomic rank that pertain to that same taxon. For example, *Archelaphe bella*, discovered and described as a distinct species by Stanley in 1917, yet four years later, in 1921, Wall described the species *Coluber leonardi*, not realizing this species was actually the same species as *Archelaphe bella.* This means *C. leonardi* is a synonym of *Archelaphe bella*, in which case the first name described takes precedence over the one that came later, as the "new" species is realized to in fact not be new / （分类学中的）同物异名，亦称同种异名。同一生物分类单元先后被给予两个或两个以上不同的学名，这些名称虽异，但实指同物。例如，Stanley于1917年发现并将"*Archelaphe bella*（方花蛇）"描述为一个独特的物种。四年后的1921年，Wall对物种"*Coluber leonardi*"进行描述。当时，Wall并未意识到这一物种实际与"*Archelaphe bella*"为同一物种。根据国际动物命名法规的"优先权律"，"*Coluber leonardi*"为"*Archelaphe bella*"的同物异名。

Taxon: A scientifically defined unit of more or less narrowly related organisms. Plural: taxa / 分类单元（复数形式：taxa）。

Terra typica: Lat. for "type locality." The location where the type specimen was found / 拉丁语，意为"模式产地（type locality）"，即模式标本的采集地。

Terrestrial: Living on the ground / 地栖的。

Thanatosis: Defensive behavior in which a prey animal feigns death. It is usually employed when escape is impossible. The technique is effective in predators that only eat prey they have killed. The simulated death posture is maintained only for a short period and then the animal 'recovers' / 装死、假死。用于迷惑捕食者的一种自卫手段。通常在逃脱无望的情况下采取该策略，对只吃自捕猎物的掠食者尤为奏效。

TL: Total length (tip of the snout, to the tip of the tail) / 全长（本书中指某蛇从吻端至尾尖的长度）。

Tubercles: Knob-like projection / 瘤状突起。

Type specimen: Specimen after which a new species was described; this is usually the first representation of that species. See **holotype** / 模式标本。某新物种据以命名、描述的那一份标本。

Vent: Another name for the **Cloaca**. The vent, or cloaca, is located at the anal plate / 泄殖腔，与"cloaca"一词意思相同。

Ventral; venter: Relating to the underside of the body; the belly; in opposition to **Dorsal** / 腹部的；腹部。本书中也指腹鳞。"Dorsal"的反义词。

Vertebral: Any one of the row of scales lying on the mid-dorsal line of the body / 本书中指脊鳞。蛇背正中的一行背鳞叫做脊鳞。

Viviparous: Live-bearing form of reproduction when the mother gives birth to live young. See **ovoviparous** / 胎生的。

Xantho-: Gr. *xanthos* = yellow / 希腊语源前缀，源自希腊语中的"*xanthos*"一词，意为"黄色，黄色的"。

Xiang Gang: The Mandarin (aka, mainland) name for "Hong Kong." Hong Kong is the Cantonese pronunciation / 香港。

Xizang: Province in southwestern China. Most foreigners know this province pronounced as Tibet. Not to be confused with Xinjiang, which is the northwestern most province / 西藏。

The genus *Archelaphe* / 方花蛇属

This is a monotypic genus.

Etymology: The generic name is derived from the Greek word *archaios*, meaning old or archaic. This refers to the basal position of the *Archelaphe* clade in the phylogenetic tree of *Elaphe*. The species was formerly classified in the genus *Maculophis*, which, when translated, means spotted (*maculos*) snake (*ophis*) and is referring to the blotched dorsum.

本属为单型属，即属下仅方花蛇（*Archelaphe bella*）一个物种。

属名词源：源于希腊语中的"*archaios*"一词，意为"古老的"。该属名同方花蛇属进化支位于锦蛇属（*Elaphe*）进化树的基部有关。作为该属下唯一的物种，方花蛇之前曾被归入丽斑蛇属（*Maculophis*）。"*maculos*"意为"有斑点的"，"*ophis*"意为"蛇"。"*Maculophis*"这一属名意指属下蛇种体背的斑纹。

General captive care / 人工饲养概述：

If available, captive-bred specimens should be acquired. Wild-caught or farm-raised specimens typically have parasites, are malnourished, and are heavily stressed, resulting in a weakened immune system. If working with wild-caught or farm-raised individuals, a qualified veterinarian should conduct a fecal and lung wash to test for parasites. Animals should be quarantined according to the protocols set out by the facility or the keeper, but an approximate quarantine period of 3 months to 1 year is advised to prevent the spread of any potential diseases or parasites.

野外捕获个体及部分养殖场所提供的个体或因寄生虫感染、营养不良、应激反应等因素存在免疫力低下的隐患。若购入以上个体，建议联系有资质的兽医对个体的呼吸道和消化道样本进行寄生虫检测。若可能，应从可靠渠道获取人工繁殖个体。新入手的个体需实行隔离，隔离时长可根据提供方给出的建议。一般建议隔离3个月至1年。

This genus is fossorial and should be treated as such in captivity with a deep layer of leaf litter and/or mulch for the species to burrow under. Adult and juvenile specimens should be offered a minimum terrarium of 61 x 41 x 16 cm. Hatchlings can be housed in a plastic shoebox or commercialized snake breeding containers measuring 31 x 19 x 10 cm for their first two years of life, after which they should be provided with a larger enclosure. Paper towels are used as the substrate for hatchlings, as this species has a very strong feeding response and is used to avoid the accidental ingestion of substrate material. Hatchling setups should also include a moist hide which contains sphagnum moss, which is commonly used when the juveniles are opaque. Individuals can be found buried in the substrate with a temperature range of 21-25℃ and relative humidity of 60-70%. With weather patterns, these animals are cycled in the fall, with temperatures gradually cooling to stimulate brumation and follicular development in females. These animals are maintained at temperatures of 10-15℃ for 3-4 months. In captivity, we have found that males' fertility drops as temperature increases, and males should be introduced in the spring after the animals have emerged from brumation before feeding. This behavior has led to the best results in breeding and fertility in the species.

人工饲养下，本属的蛇种有钻入底材的习性，故底材的铺设应达到一定的厚度。底材可选用落叶、树皮或木屑。成体及小亚成体（本书对“成体”、“亚成体”、“小亚成体”及“幼体”的定义详见“术语与缩写”章节中的解释）的饲养容器应不小于61 x 41 x 16厘米。幼体可饲养在普通塑料盒或专业饲养箱内，空间大小建议在31 x 19 x 10厘米。不过随着幼体的成长，两年后应考虑为其提供更大的饲养空间。考虑到本属蛇种有较强的摄食反应，为避免误食底材，幼体的底材建议选用厨房纸。幼体的环境中应放置加湿躲避穴，可在其中填充泥炭藓用于提高湿度。小亚成体蜕皮时也常使用该类躲避穴。当环境温度达到21～25℃，相对湿度介于60～70%，个体或钻入底材之中。为模拟本属蛇种在自然环境中的节律，饲主应在秋季逐步降低饲养环境的温度以刺激个体的冬化行为以及雌性卵泡的发育。冬化时的温度应保持在10～15℃，为期3～4个月。人工饲养下，作者发现受精率会随着温度的升高而下降。冬化后，公母应在初春首次喂食前置于同一饲养环境中以便交配的进行。此举可最大程度保证繁殖的成功。

Egg Incubation: Eggs of this genus are housed in commercially available incubators set to 24℃ or room temperature and measuring a minimum of 25 x 25 x 15 cm. Incubating at room temperature will result in a hatching delay by approximately 10-15 days. The containers must be well ventilated with holes on all sides to allow for fresh air flow and remove excessive moisture. Excessive moisture and improper ventilation can create issues for developing embryos/eggs, as mold and fungus can grow on the egg's surface. Eggs can be incubated upon a variety of media, but preference is vermiculite and perlite. Both substrates retain moisture and aid in maintaining humidity. Still, both media can dry out and may require additional water if necessary. To achieve proper moisture for incubating, dry media should be measured to have a 5 cm depth within the container. A 2:1 ratio of substrate to water by weight is normally sufficient to ensure proper moisture and humidity during incubation. One can check the appropriate balance by squeezing the mixture by hand. Expressing water means the water content is too high. After achieving the proper moisture, the media, followed by the eggs, can be placed in the incubator. When eggs are laid, it is best to gently separate eggs to allow for adequate airflow within the incubation chamber. Eggs are placed atop the substrate or partially sunk into the media to hinder the egg's movement when other eggs are hatching.

卵的孵化：使用专业孵化器以24℃或室温进行孵化皆可使本属蛇种的幼体顺利出壳，但室温下的孵化期要多10～15天。孵化器的尺寸建议不小于25 x 25 x 15厘米。盛放卵的孵化盒四周均应有孔，以保证空气的流通及多余湿气的排出。湿度过高或通风不畅将对卵的发育造成影响，易令卵壳表面滋生真菌或霉菌。孵化介质有多种选择，推荐使用蛭石或珍珠岩。两者皆有不俗的锁水性，有利于湿度的保持。然而，两种介质也可能在孵化的过程中变得过干。若有必要，需另行补水。为确保孵化过程中湿度始终处于理想水平，孵化介质(干燥状态下)在容器内的铺设厚度需达到5厘米。然后，水和孵化介质在重量上以1:2进行配比。一种检验配比是否合适的有效方法是用手捏兑水后的孵化介质。若有水渗出，则表明水份过高。放置卵时，建议互相之间保持一定的距离，以促进孵化盒内空气的流通。卵可放置在孵化介质表面或半埋于孵化介质之中。若采取半埋的方式则可限制卵的移动，从而避免对孵化造成影响。一切准备妥帖之后，孵化盒放入孵化器中。

Scientific name / 学名: *Archelaphe* [formerly *Maculophis*] *bella* (Stanley, 1917)

English name / 英文名: Elegant ratsnake, Stanley's ratsnake, Fanghua ratsnake

Chinese name / 中文名: 方花蛇 / fāng huā shé

Etymology / 种名词源: *bella* is Latin for "beautiful." The subspecies *chapaensis* is named after the type locality for that subspecies / "*bella*"，拉丁语，意为"美丽的"。*Archelaphe b. chapaensis* 亚种根据其模式产地——越南北干省沙坝村（Chapa village）来命名。

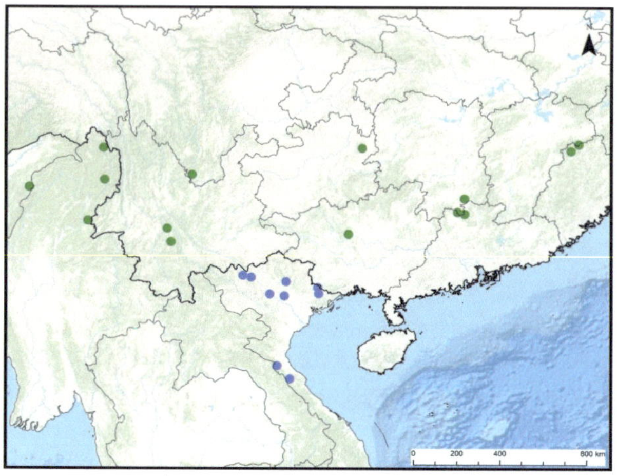

green = *A. b. bella*; blue = *A. b. chapaensis*

Terra typica / 模式产地: Guadun village, Wuyishan, Fujian, China (N27.735249° E117.638055°)/ 中国，福建省，武夷山市，挂墩村

Length / 体长: Sexual maturity is reached at 550 mm (males) and 700 mm (females); hatchlings emerge at 200-250 mm. Maximum size for males 1043 mm, 134 mm tail; for females 975 mm, 125 mm tail / 雄性体长550毫米时达到性成熟，雌性体长700毫米时达到性成熟。刚孵化的幼蛇体长200～250毫米。已知最大雄性个体全长1043毫米，其中尾长134毫米；已知最大雌性个体全长975毫米，其中尾长125毫米。

Description: The cephalic scales are as follows: supralabials 6-7, with the 3rd and 4th entering the orbit; infralabials 9, the first 5 in contact with chin shields; loreal absent; preocular 1, postoculars 2; temporals 1+2; dorsal scales in 19:19:17 rows, smooth, without apical pits. Ventrals 209-229; subcaudals 51-54, paired; anal plate divided. Head short, barely distinct from neck; snout rounded. Body cylindrical. Color above variable depending on locality and subspecies. Juveniles are more vibrantly colored than adults. Red dorsum with black and yellow bands or blotches. Yellowish white below, with a checkerboard pattern on the ventral scales. Head with Y-shaped or V-shaped markings, with the two arms ending above the eye. Juveniles resemble adults but are usually more vibrant in coloration.

Left: *A. b. bella* ; Right: *A. b. chapaensis* (Matthew Most)

外观描述：鳞被：上唇鳞6（2-2-2式）或7（2-2-3式）；下唇鳞9，前5接颌片；无颊鳞；眶前鳞1；眶后鳞2；颞鳞1+2；背鳞19-19-17行，鳞片平滑，无端窝；腹鳞209～229；尾下鳞51～54对；肛鳞二分。头小，与颈区分不明显。吻端圆钝。体圆柱形。体背的颜色因地域及亚种的不同而不同。小亚成体在外观上与成体相似，但体色更为艳丽。本种体背红色，具黑、黄斑纹。腹面黄白色，散有黑色方斑。头背具"Y"或"V"形斑纹。

Distribution: Within Greater China, the species is found in Fujian, Jiangxi, Guangdong, Guangxi, Guizhou, Hunan, Sichuan, and Yunnan. Outside of Greater China, the species is found in India, Myanmar, and Vietnam. Found from 1000-3000 m. Not recorded from Laos yet, but very likely present in northern Laos.

地理分布：在中国，分布于福建、江西、广东、广西、贵州、湖南、四川和云南。中国以外，分布于印度、缅甸和越南。垂直分布1000～3000米。虽尚未有于老挝的分布记录，但该国北部很可能有本种分布。

Habits and habitats: Very poorly known. The species is seemingly very cryptic, as specimens are infrequently encountered. The species is known to be an inhabitant of mountain forests, often associated with rivers and big boulders. Most specimens remain quietly coiled or try to flee when disturbed. Still, some individuals may be quite vicious, flattening, jerking and twitching their heads, vibrating their tails, striking and biting, and spraying musk and cloacal contents when picked up. Defensive behavior is temperature-dependent—the colder the snake, the slower it reacts.

习性与栖息地：对此知之甚少。生性极为隐秘，不易被发现。常栖息于山林之中，周围多有河流和巨石。如遇惊扰，大部分个体原地盘曲或设法逃离。然而，也有部分个体或较为凶猛，感知威胁后常使身体呈扁平状，摆动头部，抖动尾部并扑咬。若被抓取还会分泌恶臭液体。防御行为取决于温度。温度越低，反应越慢。

Diet and feeding habits: In captivity, the species will take small mammals such as rodents. In the wild, this species likely preys on rodents' nests as juvenile and adult animals in captivity prefer multiple smaller rodent prey items. Captive hatchlings feed on newborn mice readily without scenting.

食性与喂养：人工饲养下，以啮齿类等小型哺乳动物为食。根据人工饲养下的食性，推测本种在自然环境中捕食啮齿类刚出生的幼崽。人工繁殖的幼蛇可直接进食在气味上未经特殊处理的乳鼠。

Habitat shot of *A. b. bella* from northern Guangdong.

Captive care: See the general captive care section at the introduction of this genus. Hatchling setups also include a moist hide which contains sphagnum moss, which is commonly used when the juveniles are opaque. Adults and juveniles are housed in enclosures that are 61 x 41 x 16 cm. Adult and juvenile specimens prefer a deep layer of mulch and leaf litter as substrate. Individuals can be found buried in the substrate with a temperature range of 21-25°C and relative humidity of 60-70%. In captivity, this species has been recorded to have a life span of 10-15 years.

人工饲养：具体情况请参阅本属介绍章节中的"人工饲养概述"版块。幼体的环境中应放置加湿躲避穴，可填充泥炭藓用于提高湿度。小亚成体蜕皮时也常使用该类躲避穴。成体及小亚成体的饲养容器应不小于61 x 41 x 16厘米。成体与小亚成体更偏爱以树皮、木屑或落叶作为底材，底材的铺设应达到一定厚度。当环境温度达到21～25°C，环境湿度介于60～70%，方花蛇或钻入底材之中。据记载，人工饲养下本种的寿命为10～15年。

A. b. chapaensis hatching (Matthew Most).

Reproduction: Oviparous. *Archelaphe* reaches sexual maturity in 3-4 years. In captivity, specimens are provided with a brumation period of 10-15°C for three months. After brumation, specimens are paired immediately in early spring before they have fed for copulation. Copulation can last several days, but male fertility will decrease as temperatures increase. The gestation period for *Archelaphe* is 60 days, with eggs being laid in May and June. Clutch size ranges from 5-12 eggs with incubation of eggs at 23-25°C for 60-70 days. Females can produce a second clutch within the same year, as has been observed in private collections.

繁殖：卵生。于3～4年达到性成熟。人工饲养下，个体应在10～15°C下经历三个月的冬化期。冬化后，公母个体应在初春首次喂食前置于同一饲养环境中以便交配的进行。交配可持续数日，但受精率会随着温度的升高而下降。母蛇怀卵约60天。5～6月产卵。一窝卵5～12枚。孵化温度23～25°C，孵化期60～70天。人工饲养下，本种曾有一年产卵两窝的记录。

Taxonomy and nomenclature: Schultz (2000) showed that *Elaphe leonardi* (Wall, 1921) is a synonym of *Coronella bella* (Stanley, 1917), now *Archelaphe bella*.

Archelaphe b. bella (Stanley, 1917): China and Myanmar

Archelaphe b. chapaensis (Bourret, 1934): N Vietnam (Bac-Kan), Myanmar, China (W Yunnan); Type locality: near Chapa village (= Sa Pa), Bac Kan province, N Vietnam.

Left: *A. b. bella* specimen from Longchuan County, Yunnan (Gernot Vogel).
Right: *A. b. bella* specimen, found dead on the road, in northern Guangdong (T.P. Chen 陈太平).

分类与命名：Schultz（2000）指出：*Elaphe leonardi* (Wall, 1921)实为*Coronella bella* (Stanley, 1917)的同物异名。而*Coronella bella*就是现如今的*Archelaphe bella*。

Archelaphe b. bella (Stanley, 1917)：分布于中国和缅甸。

Archelaphe b. chapaensis (Bourret, 1934)：分布于越南北部（北干省）、缅甸、中国（云南西部）。模式产地：越南，北干省，沙坝村附近。

Conservation Status: IUCN considers this species Least Concern (LC).

保护状况：本种在《濒危物种红色名录》中被定为"无危"等级。

A juvenile *A. b. chapaensis* specimen (Matthew Most).

The genus *Coelognathus* / 颌腔蛇属

This genus consists of seven species, only one of which is found within China. This genus was split from *Elaphe* in 2001 by Helfenberger, revalidating the genus *Coelognathus* that was proposed by Leopold Fitzinger in 1843. Members of the genus consist of very active and alert species.

本属下共有7个物种，仅一种（三索锦蛇 / *Coelognathus radiatus*）在中国有所分布。2001年，Helfenberger将本属从锦蛇属中拆分，并恢复使用Leopold Fitzinger于1843提出的颌腔蛇属。本属下的蛇种较为活跃、机警。

Etymology: From the Latinized form of Greek *koilos,* meaning "hollow, concave." And *–gnathos* (Greek: *gnatos*) meaning "jaw." Fitzinger (1843) did not explain the etymology.

属名词源：源自希腊语，"*koilos*" 意为 "凹陷的"，"*gnatos*" 意为 "颌"。Fitzinger (1843) 对此命名没有给出解释。

General captive care / 人工饲养概述:

If available, captive-bred specimens should be acquired. Wild-caught or farm-raised specimens typically have parasites, are malnourished, and are heavily stressed, resulting in a weakened immune system. If working with wild-caught or farm-raised individuals, a qualified veterinarian should conduct a fecal and lung wash to test for parasites. Animals should be quarantined according to the protocols set out by the facility or the keeper, but an approximate quarantine period of 3 months to 1 year is advised to prevent the spread of any potential diseases or parasites.

野外捕获个体及部分养殖场所提供的个体或因寄生虫感染、营养不良、应激反应等因素存在免疫力低下的隐患。若购入以上个体，建议联系有资质的兽医对个体的呼吸道和消化道样本进行寄生虫检测。若可能，应从可靠渠道获取人工繁殖个体。新入手的个体需实行隔离，隔离时长可根据提供方给出的建议。一般建议隔离3个月至1年。

This genus is very active and uses every space of their enclosure, and does well in both an arboreal and terrestrial setup. Individuals will use vertical climbing space if provided with multiple basking areas. This species is kept at temperatures that range from 24-30° C with a relative humidity of 50-60%. Adult specimens and juveniles should be offered a minimum terrarium of 86 x 46 x 16 cm. Adult and juvenile specimens are kept on aspen bedding as a substrate, with a moist hide being provided to females as specimens will lay eggs throughout the year. Unless breeding is desired, males and females should be kept separate, as males will court females throughout the year, resulting in multiple clutches being laid. If females cannot maintain proper weight, they will degrade in captivity quickly and eventually die due to poor health. Hatchlings can be housed in a plastic shoebox or commercialized snake breeding containers measuring 31 x 19 x 10 cm. Paper towels are used as the substrate for hatchlings, as this species has a very strong feeding response and is used to avoid accidental ingestion of substrate debris.

本属的蛇种较为活跃。人工饲养下会充分利用饲养环境中的每寸空间。环境布置方面，无论是以树栖还是地栖为主，本属的蛇种皆可很好适应。若饲养场地设置有多个晒点，个体会充分利用环境中的垂直空间。饲养环境的温度应为24～30℃，相对湿度介于50～60%。成体及小亚成体的饲养容器应不小于86 x 46 x 16厘米。成体和小亚成体的底材可选用山杨木屑。雌性可全年产卵，建议提供加湿躲避穴。雄性全年都会有求偶行为，导致雌性一年内数次产卵。若雌蛇无法维持适当的体重，这或将导致其健康状况迅速恶化甚至死亡。故若非出于繁殖目的，公母个体应分开饲养。幼体可饲养在普通塑料盒或专业饲养箱内，空间大小建议在31 x 19 x 10厘米。考虑到本属蛇种有较强的摄食反应，为避免误食底材，幼体的底材建议选用厨房纸。

Egg Incubation: Eggs of this genus are housed in commercially available incubators set to 27-28°C or room temperature and measuring a minimum of 25 x 25 x 15 cm. Incubating at room temperature will result in a hatching delay by approximately 10-15 days. The containers must be well ventilated with holes on all sides to allow for fresh air flow and remove excessive moisture. Excessive moisture and improper ventilation can create issues for developing embryos/eggs, as mold and fungus can grow on the egg's surface. Eggs can be incubated upon a variety of media, but preference is vermiculite and perlite. Both substrates retain moisture and aid in maintaining humidity. Still, both media can dry out and may require additional water if necessary. To achieve proper moisture for incubating, dry media should be measured to have a 5 cm depth within the container. A 2:1 ratio of substrate to water by weight is normally sufficient to ensure proper moisture and humidity during incubation. One can check the appropriate balance by squeezing the mixture by hand. Expressing water means the water content is too high. After achieving the proper moisture, the media, followed by the eggs, can be placed in the incubator. When eggs are laid, it is best to gently separate eggs to allow for adequate airflow within the incubation chamber. Eggs are placed atop the substrate or partially sunk into the media to hinder the egg's movement when other eggs are hatching.

卵的孵化：使用专业孵化器以27～28℃或室温进行孵化皆可使幼体顺利出壳，但室温下的孵化期要多10～15天。孵化器的尺寸建议不小于25 x 25 x 15厘米。盛放卵的孵化盒四周均应有孔，以保证空气的流通及多余湿气的排出。湿度过高或通风不畅将对卵的发育造成影响，易令卵壳表面滋生真菌或霉菌。孵化介质有多种选择，推荐使用蛭石或珍珠岩。两者皆有不俗的锁水性，有利于湿度的保持。然而，两种介质也可能在孵化的过程中变得过干。若有必要，需另行补水。为确保孵化过程中湿度始终处于理想水平，孵化介质（干燥状态下）在容器内的铺设厚度需达到5厘米。然后，水和孵化介质在重量上以1:2进行配比。一种检验配比是否合适的有效方法是用手捏兑水后的孵化介质。若有水渗出，则表明水份过高。放置卵时，建议互相之间保持一定的距离，以促进孵化盒内空气的流通。卵可放置在孵化介质表面或半埋于孵化介质之中。若采取半埋的方式则可限制卵的移动，从而避免对孵化造成影响。一切准备妥帖之后，孵化盒放入孵化器中。

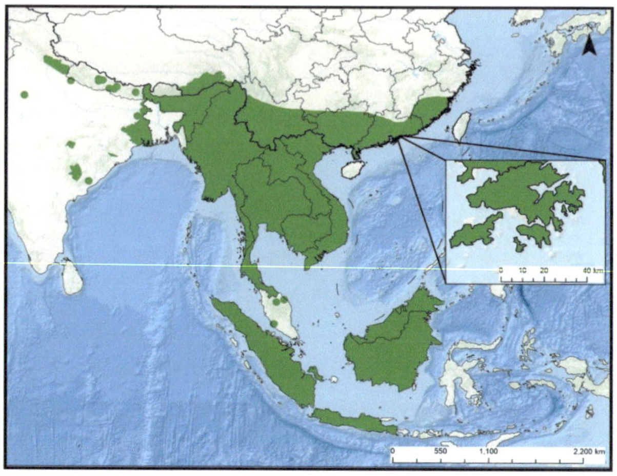

Scientific name / 学名: *Coelognathus* [formerly *Elaphe*] *radiatus* (Boie, 1827)

English name / 英文名: Radiated ratsnake, Copperhead racer, Copperhead ratsnake, Copperheaded trinket snake

Chinese name / 中文名: 三索锦蛇 / sān suǒ jǐn shé

Etymology / 种名词源: Latin, *radius*: for radial, referring to the radiate line pattern around the eyes / 源自拉丁语中的"*radius*"一词，意为"辐射状的"。该种名意指三索锦蛇眼周围的辐射状细纹。

Terra typica / 模式产地: "Java, Indonesia" by type locality restriction. Further restricted to the region of Mt. Pangerango and Mt. Salak, West Java, Indonesia *fide* Brongersman (1948) (circa S6.72° E106.73°) / 印度尼西亚，爪哇岛。据Brongersman（1948）提供的可靠信息，本种模式产地限于爪哇岛西部的庞朗奥火山和萨拉火山区域。

Length / 体长: Maturity is attained at 900 mm; hatchlings emerge around 250-340 mm, and the species reaches a maximum size around 2100 mm / 体长900毫米时达到性成熟。刚孵化的幼蛇体长250～340毫米。本种已知的最大体长在2100毫米左右。

Description: The cephalic scales are as follows: Supralabials 9 rarely 8; with 4-6 or 3-5 in contact with the orbit; infralabials 9-11; preocular 1; postocular 2; temporals usually 2. Dorsal scales in rows of 19: 19-21: 17 anterior to posterior, respectively. Ventrals number between 207 -250 with a strong lateral keel; anal entire; subcaudals 80-108, paired. Head is copper-brown, with a narrow black line radiating from the eyes downward, a second radiating backward, and a third radiating at an angle diagonally in-between (uppermost streak joins black line on nape). The transverse line meets a thin black collar line. The body, which is highly compressed laterally in the area immediately behind the head and neck, is grayish-brown, fawn, yellowish-brown, or gray, with four black stripes running down the dorsum on each side. These black stripes gradually narrow and turn down the dorsum on each side. The tail is light brown. Adults have smooth or very weakly keeled scales. The upper stripes are broad, with the outer pair usually broken into elongated spots. The ventral is uniformly yellow or gray. Juveniles resemble adults.

外观描述：鳞被：上唇鳞9（3-3-3、2-3-4式），少数8（3-3-2、2-3-3式）；下唇鳞9～11；眶前鳞1；眶后鳞2；颞鳞通常2；背鳞颈部19行，中段19～21行，肛前17行；腹鳞207～250，侧棱明显；肛鳞完整；尾下鳞80～108对。头部呈铜棕色。眼周围有三条放射状黑色细纹；一条垂直于眼向下，另一条由眼向后延伸，第三条斜切于上述两条之间。其中，眼后方的细纹与枕部的黑色细横纹相切。颈部之后的躯体侧扁。背面浅黄褐、灰褐或灰色，两侧各有两条黑色纵纹，纵纹向体后延展，逐渐变细；靠近正背的一对纵纹明显更宽，下方的一对纵纹常不规则断裂。尾部为浅棕色。腹面为黄或灰色。成体背鳞平滑或微弱起棱。小亚成体与成体在外观上相似。

Distribution: *C. radiatus* is the most widely distributed member of the genus. Within Greater China, this species is found in Hong Kong, Guangdong, Guangxi, Guizhou, and Yunnan. Elsewhere, it is found in Bangladesh, Borneo, Cambodia, India, Indonesia, Laos, Malaysia, Myanmar, Nepal, Singapore, Thailand, and Vietnam. Found in elevations ranging from 0-1800 m.

地理分布：三索锦蛇为其属下分布最广的蛇种。在中国，分布于香港、广东、广西、贵州和云南。中国以外，分布于印度尼西亚、马来西亚、婆罗洲（分属印度尼西亚、马来西亚和文莱三国）、新加坡、缅甸、泰国、老挝、柬埔寨、越南、印度、孟加拉国和尼泊尔。垂直分布0～1800米。

Habits and habitats: This species is primarily diurnal. Though terrestrial, this species is an excellent climber and swimmer as well. Frith (1977) even reported swimming activities in the open sea. Commonly found in alluvial grassland, plantations, agricultural, urban, and open areas near forests. When encountered in the open, *C. radiatus* quickly crawls away and sometimes even climbs into trees or bushes to evade capture.

习性与栖息地：白昼活动为主。常见于森林周边的农田、城市和空旷地带。作为一个地栖蛇种，三索锦蛇也十分擅长攀附和游泳。Frith（1977）曾报道本种于海域活动。若在野外相遇，三索锦蛇会迅速爬开，有时甚至攀援上树或躲入灌木。

A young specimen from Jangcheng, Yunnan (Gernot Vogel).

General habitat shot of *C. radiatus*. This species lives in a wide range of habitats.

If cornered or prevented from fleeing, it will face the attacker, throw its forebody into several S-coils, puff up with air in an attempt to look larger, vertically spread its neck, gape, hiss, and strike repeatedly. When handled, it thrashes about, bites (and often chews), and sprays musk. However, if continually harassed, some specimens are known to exhibit thanatosis (the act of playing dead). The snake will stay in this condition until the aggressor loses interest in the animal and moves on.

Above and right: Threat displays from an adult specimen in Hong Kong (Left: Dan Rosenberg; Right: Kevin Caldwell).

Below: *C. radiatus* specimen playing dead (thanatosis) until its aggressor loses interest in it (Adam Francis).

若无法逃脱，则会呈迎击姿态。此时，其常侧扁颈部，体前段呈“S”形卷曲鼓起，张嘴发出嘶嘶声并反复发动攻击。若被抓取，会剧烈扭动身体并试图扑咬，还会分泌恶臭液体。如遇反复滋扰，个别还会采取装死的策略。

Diet and feeding habits: This species kills prey via constriction. Adults feed primarily on rodents and birds, juveniles primarily on lizards. Also known to prey on bats, especially around human dwellings. Occasionally the species is known to prey on frogs.

食性与喂养：有绞杀猎物的习性。成体主要以啮齿类和鸟类为食，小亚成体主要以蜥蜴为食。已知还会捕食蝙蝠和蛙类。

Captive care: See the genus introduction for general care. In captivity, *C. radiatus* has been recorded to have a life span of 15-20 years.

人工饲养：具体情况请参阅本属介绍章节中的"人工饲养概述"版块。据记载，人工饲养下本种的寿命为15～20年。

Reproduction: Oviparous. *Coelognathus* can reach sexual maturity in 2-3 years. In captivity, brumation is not required for captive reproduction as this species can and will breed throughout the year. In the wild, this species will lay multiple clutches per year, relative to seasonal change, with most clutches coinciding with the warm months of the year. The gestation period for *Coelognathus* is 40-45 days, with eggs being incubated at temperatures of 26-29°C for 60-95 days. The clutch size ranges from 6-13 eggs. Females are also capable of producing multiple clutches a year, upwards of 9 clutches per year.

A juvenile from Hong Kong, in threat display, gaping its mouth in an attempt to scare away its antagonist (Dan Rosenberg).

繁殖：卵生。可于2～3年达到性成熟。人工饲养下全年皆可繁殖且无需冬化。自然环境中，本种一年产卵数窝，但多集中于一年中相对温暖的月份。母蛇怀卵40～45天。一窝卵6～13枚。孵化温度26～29℃，孵化期60～95天。母蛇一年最多可产卵九窝。

Taxonomy and nomenclature: Recent molecular work removed this species from the genus *Elaphe* and placed it in the genus *Coelognathus.* There are no subspecies recognized.

分类与命名：基于最近的分子生物学研究，三索锦蛇被从锦蛇属中移出，归隶颌腔蛇属。本种目前无公认的亚种。

Conservation Status: The conservation status of this species is Least Concern (LC) according to the IUCN.

保护状况：本种在《濒危物种红色名录》中被定为“无危”等级。

The genus *Elaphe* / 锦蛇属

Most species of this genus are large animals, usually more than 1 m in length and usually not greater than 2 m, with a few notable exceptions. All members of the genus are constrictors. Most are diurnal, but several species will also become active at night, especially during the hotter summer months. Generally speaking, the genus is both terrestrial and arboreal.

本属下大多数蛇种体型较大，体长通常超过1米但不足2米，但也有部分特例存在。属下所有成员皆有绞杀猎物的习性。大多数蛇种白昼活动，但也有部分于夜间保持活跃，尤其是在温度较高的夏季。一般来说，锦蛇属的蛇种既有地栖又有树栖习性。

Etymology: *Elaphe* is from the Greek *elaphos*, meaning deer or stag. There are two possible explanations: the type species for the genus is *Elaphe sauromates* and could be referencing the antler-like markings on this species' head. Another explanation is presumed to be about swiftness, as some members of the genus are quick to flee upon discovery.

属名词源：源自希腊语中的“*elaphos*”一词，意为“鹿或成年雄鹿”。对于该属名，存在两种可能的解释。该属的模式种为东四线锦蛇（*Elaphe sauromates*），该蛇的头部有酷似鹿角的斑纹。另一种解释或跟本属部分蛇种的敏捷性有关：它们被发现后往往能快速逃离，这点也和鹿的习性颇为相像。

Elaphe sauromates, the type species for which *Elaphe* was described upon (Dick Barlett).

Taxonomy notes: As recently as the 1990s, nearly all species covered in this book (those known to science at the time) were assigned to the genus *Elaphe*. In the early 2000s, several of these Asian members were reassigned and broken into *Archelaphe, Coelognathus, Euprepiophis, Gonyosoma, Maculophis, Oreocryptophis, Orthriophis,* and *Rhadinophis.*

分类注释：上世纪90年代，本书所介绍的蛇种在当时几乎全都归隶锦蛇属（*Elaphe*）。21世纪前几年，它们中的一些或改隶、或被拆分成方花蛇属（*Archelaphe*）、颌腔蛇属（*Coelognathus*）、玉斑蛇属（*Euprepiophis*）、树栖锦蛇属（*Gonyosoma*）、丽斑蛇属（*Maculophis*）、紫灰蛇属（*Oreocryptophis*）、晨蛇属（*Orthriophis*）和绿蛇属（*Rhadinophis*）。

Some of these arrangements were well received by most experts familiar with the species in question, and others received a lot of opposition. For example, *E. cantoris, E. hodgsonii, E. moellendorffi,* and *E. taeniura* were all reassigned to *Orthriophis* in 2010, then, just a few years later, were reassigned back into *Elaphe.* Because such back and forth reassessments can lead to confusion, I try to include the genus and the species it was formerly aligned with most recently for species subjected to such situations. Given the trajectory of recent molecular studies, they may end up back in the said genus.

以上这些变化中，有些获得了绝大多数专家的支持，有些却遭到了反对。例如，坎氏锦蛇（*Elaphe cantoris*）、南峰锦蛇（*Elaphe hodgsonii*）、百花锦蛇（*Elaphe moellendorffi*）和黑眉锦蛇（*Elaphe taeniura*）在2010年改隶晨蛇属。但短短数年后，四者又重新回归锦蛇属。考虑到如此反复极易引起混淆，对于这些蛇种，它们目前所在和上一次所在的属于本书中都将有所体现。或许在将来，相关的生物分子学研究又将使它们的归属出现变化。

General captive care / 人工饲养概述:

If available, captive-bred specimens should be acquired. Wild-caught or farm-raised specimens typically have parasites, are malnourished, and are heavily stressed, resulting in a weakened immune system. If working with wild-caught or farm-raised individuals, a qualified veterinarian should conduct a fecal and lung wash to test for parasites. Animals should be quarantined according to the protocols set out by the facility or the keeper, but an approximate quarantine period of 3 months to 1 year is advised to prevent the spread of any potential diseases or parasites.

野外捕获个体及部分养殖场所提供的个体或因寄生虫感染、营养不良、应激反应等因素存在免疫力低下的隐患。若购入以上个体，建议联系有资质的兽医对个体的呼吸道和消化道样本进行寄生虫检测。若可能，应从可靠渠道获取人工繁殖个体。新入手的个体需实行隔离，隔离时长可根据提供方给出的建议。一般建议隔离3个月至1年。

Egg Incubation: Eggs of this genus are housed in commercially available incubators set to 27-28°C or room temperature and measuring a minimum of 20 x 20 x 15 cm. However, depending on the number of eggs, this size may be increased or decreased. For example, eggs of *Elaphe anomala, E. carinata,* and *E. schrenkii* are much larger than that of *E. bimaculata, E. davidi, E. dione,* and *E. zoigeensis.* Incubating at room temperature will result in a hatching delay by approximately 10-15 days. The containers must be well ventilated with holes on all sides to allow for fresh air flow and remove excessive moisture. Excessive moisture and improper ventilation can create issues for developing embryos/eggs, as mold and fungus can grow on the egg's surface. Eggs can be incubated upon a variety of media, but preference is vermiculite and perlite.

卵的孵化：使用专业孵化器以27～28℃或室温进行孵化皆可使本属蛇种的幼体顺利出壳，但室温下的孵化期要多10～15天。孵化器的尺寸建议不小于20 x 20 x 15厘米。考虑到赤峰锦蛇（*Elaphe anomala*）、王锦蛇（*Elaphe carinata*）和棕黑锦蛇（*Elaphe schrenkii*）的卵比双斑锦蛇（*Elaphe bimaculata*）、团花锦蛇（*Elaphe davidi*）、白条锦蛇（*Elaphe dione*）和若尔盖锦蛇（*Elaphe zoigeensis*）的卵大得多，孵化器的具体大小还可根据卵的数量灵活调整。盛放卵的孵化盒四周均应有孔，以保证空气的流通及多余湿气的排出。湿度过高或通风不畅将对卵的发育造成影响，易令卵壳表面滋生真菌或霉菌。孵化介质有多种选择，推荐使用蛭石或珍珠岩。

Both substrates retain moisture and aid in maintaining humidity. Still, both media can dry out and may require additional water if necessary. To achieve proper moisture for incubating, dry media should be measured to have a 5 cm depth within the container. A 2:1 ratio of substrate to water by weight is normally sufficient to ensure proper moisture and humidity during incubation. One can check the appropriate balance by squeezing the mixture by hand. Expressing water means the water content is too high. After achieving the proper moisture, the media, followed by the eggs, can be placed in the incubator. When eggs are laid, it is best to gently separate eggs to allow for adequate airflow within the incubation chamber. Eggs are placed atop the substrate or partially sunk into the media to hinder the egg's movement when other eggs are hatching.

两者皆有不俗的锁水性，有利于湿度的保持。然而，两种介质也可能在孵化的过程中变得过干。若有必要，需另行补水。为确保孵化过程中湿度始终处于理想水平，孵化介质（干燥状态下）在容器内的铺设厚度需达到5厘米。然后，水和孵化介质在重量上以1:2进行配比。一种检验配比是否合适的有效方法是用手捏兑水后的孵化介质。若有水渗出，则表明水份过高。放置卵时，建议互相之间保持一定的距离，以促进孵化盒内空气的流通。卵可放置在孵化介质表面或半埋于孵化介质之中。若采取半埋的方式则可限制卵的移动，从而避免对孵化造成影响。一切准备妥帖之后，孵化盒放入孵化器中。

General captive care for *Elaphe bimaculata, E. davidi, E. dione,* and *E. zoigeensis:*

This section is the generalized care for the species mentioned. Please look at the Captive Care section under the species of interest for species specifics. These are the small to medium-sized *Elaphe* species. All these species do well in a setup with less moisture and more ventilation than the other species in this genus. If inadequately ventilated, these species may develop symptoms of a respiratory tract infection. Adult and juvenile specimens should be offered a minimum terrarium of 61 x 41 x 16 cm. Hatchlings can be housed in a plastic shoebox or commercialized snake breeding containers measuring 31 x 19 x 10 cm for their first two years of life, after which they should be transferred to a larger terrarium. Hatchling setups should also include a moist hide which contains sphagnum moss, which is commonly used when the juveniles are opaque. Individuals can be found buried in the substrate with a temperature range of 21-25°C and relative humidity of 40-50%. In captivity, these animals often breed in autumn. Still, it is unknown if this serves as a reproductive advantage of the species or merely a result of attraction. If kept in communal setups, you should note feeding behaviors as sexual dimorphism exists within these species, and ophiophagy has been noted in these species, with larger individuals consuming smaller individuals. The breeding in the fall season does not result in eggs during the cooler months in any of these species. These species are also cycled with weather patterns. These species are maintained at temperatures of 10-15°C for 3-4 months. To induce breeding behavior, males should be introduced to females following brumation.

双斑锦蛇、团花锦蛇、白条锦蛇、若尔盖锦蛇饲养概述：

本节是针对以上蛇种的饲养方法概述。欲知更多细节，请参阅各蛇种介绍章节中的“人工饲养”版块。这些都是中小体型的锦蛇物种。与本属其它成员相比，双斑锦蛇、团花锦蛇、白条锦蛇、若尔盖锦蛇更偏爱湿度较低且通风良好的环境。若通风不佳，则较易出现呼吸道感染症状。成体及小亚成体的饲养容器应不小于61 x 41 x 16厘米。幼体可饲养在普通塑料盒或专业饲养箱内，空间大小建议在31 x 19 x 10厘米。不过随着幼体的成长，两年后应考虑为其提供更大的饲养空间。幼体的环境中应放置加湿躲避穴，可在其中填充泥炭藓用于提高湿度。小亚成体蜕皮时也常使用该类躲避穴。当环境温度达到21～25℃，相对湿度介于40～50%，个体或钻入底材之中。人工饲养下，这些蛇种通常在秋季进行交配。对此，尚不知是某种繁殖优势亦或是信息素的作用。由于性别二态性，公母体型有较大差异，群养下需注意摄食行为。目前已发现较大个体有吞食较小个体的现象。虽然于秋季进行交配，但雌性却往往在来年的春季产卵。为模拟这些蛇种在自然环境中的节律，个体应经历为期3～4个月的冬化，冬化时的温度应保持在10～15℃。为诱发繁殖行为，公母个体应在冬化后置于同一饲养环境中。

General captive care for *Elaphe anomala, E. carinata,* and *E. schrenkii*:

This section is the generalized care for the species mentioned. Please look at the Captive Care section under the species of interest for species specifics. These are the larger *Elaphe* species and can reach 2 m or more in length as adults, requiring larger housing than smaller ones. These are all highly active species that do well in both an arboreal and terrestrial setup. Individuals will use vertical climbing space if provided with multiple basking areas. These species are kept at temperatures that range from 21-25℃ and relative humidity of 60-70%. Adult and juvenile specimens should be offered a minimum terrarium of 86 x 46 x 16 cm. Hatchlings can be housed in a plastic shoebox or commercialized snake breeding containers measuring 31 x 19 x 10 cm for the first year of life, after which they should be transferred to a larger terrarium. Paper towels are used as the substrate for hatchlings, as this species has a very strong feeding response and is used to avoid the accidental ingestion of substrate material. Hatchling setups should also include a moist hide which contains sphagnum moss, which is commonly used when the juveniles are opaque. Some neonates may not start feeding immediately after hatching and should be offered a short cooling period to stimulate the endocrine system. After a brief period of 2-3 weeks at similar temperatures offered for brumation, these individuals will start feeding. These species are also cycled with weather patterns. These animals are maintained at temperatures of 10-15℃ for 3-4 months. To induce breeding behavior, males should be introduced to females following brumation.

赤峰锦蛇、王锦蛇、棕黑锦蛇饲养概述：

本节是针对以上蛇种的饲养方法概述。欲知更多细节，请参阅各蛇种介绍章节中的“人工饲养”版块。这些都是体型较大的锦蛇物种，成年后长度可超过2米，需要较大的饲养空间。赤峰锦蛇、王锦蛇、棕黑锦蛇都是异常活跃的蛇种。饲养环境的布置无论是以树栖还是地栖为主，它们皆可很好地适应。若饲养场地设置有多个晒点，它们会充分利用环境中的垂直空间。饲养环境的温度应为21～25℃，相对湿度介于60～70%。成体及小亚成体的饲养容器应不小于86 x 46 x 16厘米。幼体可饲养在普通塑料盒或专业饲养箱内，空间大小建议在31 x 19 x 10厘米。不过随着幼体的成长，一年后应考虑为其提供更大的饲养空间。考虑到这些蛇种有较强的摄食反应，为避免误食底材，幼体的底材建议选用厨房纸。幼体的环境中应放置加湿躲避穴，可在其中填充泥炭藓用于提高湿度。小亚成体蜕皮时也常使用该类躲避穴。一些刚孵化的个体在出壳后或不会立即进食，此时应短暂降温以刺激其内分泌系统。2～3周的低温期后（温度与冬化时的温度一致），个体往往会开始进食。为模拟这些蛇种在自然环境中的节律，个体应经历为期3～4个月的冬化，冬化时的温度应保持在10～15℃。为诱发繁殖行为，公母个体应在冬化后置于同一饲养环境中。

Scientific name / 学名: *Elaphe anomala* (Boulenger, 1916)

English name / 英文名: Southern Amur ratsnake, Korean ratsnake

Chinese name / 中文名: 赤峰锦蛇 / chì fēng jǐn shé

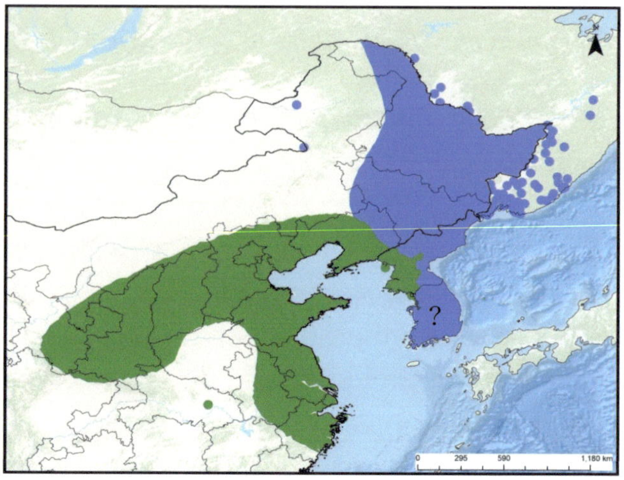

green = *E. anomala*, blue = *E. schrenkii*

Etymology / 种名词源: *Anomala* is Latin for an anomaly, referring to the strange coloration of this species compared to the nominate form, as at the time, *E. anomala* was described as a subspecies of *E. schrenkii* / "anomala", 拉丁语, 意为"反常的事物"。在本种命名时, 赤峰锦蛇仍是棕黑锦蛇的一个亚种。同棕黑锦蛇的指名亚种相比, 赤峰锦蛇的体色明显不同, 故因而得名。

Terra typica / 模式产地: Chifeng, Nei Mongol, China (N42.25° E118.883°)/ 中国, 内蒙古自治区, 赤峰市。

Length / 体长: Maturity is reached at 1100 mm; hatchlings emerge at approximately 280-450 mm. Adult males reach a maximum of up to 2070 (tail 270) mm, and females up to 2200 (tail 275) mm / 体长1100毫米时达到性成熟。刚孵化的幼蛇体长280～450毫米。已知最大雄性个体全长2070毫米, 其中尾长270毫米; 已知最大雌性个体全长2200毫米, 其中尾长275毫米。

Description: The cephalic scales are as follows: supralabials 8, rarely 7; preoculars 1 or 2; postoculars 2, rarely 1; anterior temporals 2, rarely 1; posterior temporals 1-4; scales keeled in 23 (rarely 21):23:19 (rarely 17) rows; ventrals in males 206-216, in females 218-222; paired subcaudals in males 70-77, in females 64-72; a few subcaudals sometimes single. The anal plate is divided. The species exhibits ontogenetic changes. The juvenile pattern is significantly different from the adult. Juveniles are banded with a series of dark gray, black-edged bands separated by light gray interspaces, each of which has a dark median area. Anteriorly, the dark bands are wide, the interspaces narrow. In contrast, posteriorly, bands and interspaces are about equal in width, but the former's black margins are now relatively much wider. In contrast, adults are almost uniformly brown, losing all this elaborate coloration, save for strong banding on the posterior half. Ventral usually yellowish.

Specimen from Beijing.

外观描述：上唇鳞8，少数7；眶前鳞1或2；眶后鳞2，少数1；前颞鳞2，少数1；后颞鳞1～4；背鳞起棱，23（少数21）-23-19（少数17）行；雄性腹鳞206～216，雌性腹鳞218～222；雄性尾下鳞70～77对，雌性尾下鳞64～72对；部分尾下鳞成单；肛鳞二分。小亚成体与成体在外观上有较大差异。小亚成体背面具间隔排列的深灰色和浅灰色横斑，深灰色横斑外缘镶有黑边，浅灰色横斑中心具黑线；体前段，深灰色横斑明显宽于浅灰色横斑；体后段，两者宽度大致相等，但深灰色横斑外缘的黑边明显变粗。相反，除体后段有明显的横斑外，成体赤峰锦蛇几乎通体为棕色。腹面通常为浅黄色。

Distribution: Within Greater China, this species is found in Anhui, Beijing, Gansu, Hebei, Henan, Hubei, Inner Mongolia, Jiangsu, Liaoning, Ningxia, Shaanxi, Shanxi, Shandong, Sichuan, Tianjin, and Zhejiang. It is found outside of China in the Korean peninsula, from low altitude plains up to 1800 m elev. Previous authors have also placed a disjunct population in central Hunan, but this is likely a case of mistaken identity as no specimens in recent history have been found in the province, despite several protected areas throughout Hunan conducting multiple biodiversity surveys. Due to the unsettled nature surrounding the taxonomy on *E. anomala* and *E. schrenkii*, I am providing a distribution map of both species/ subspecies. The green distribution represents what is thought to be *E. anomala;* the blue distribution represents what is thought to be *E. schrenkii*. For many years, the population in South Korea was thought to be an intergrade/ hybrid zone (depending on the stance taken on the distinction of the species/ subspecies). Recent studies have demonstrated that based on mtDNA and microsatellite data, the South Korea population aligns with the Russian ratsnake (*Elaphe schrenkii*).

地理分布：在中国，分布于安徽、北京、甘肃、河北、河南、湖北、内蒙古、江苏、辽宁、宁夏、陕西、山西、山东、四川、天津和浙江。中国以外，分布于朝鲜半岛。垂直分布可达海拔1800米。在部分文献中，本种的地理分布也包括湖南省中部地区。近年，湖南省境内的各大自然保护区纷纷对区内生物多样性展开调研，但始终未见有赤峰锦蛇的分布报道。鉴于此，本种于湖南省的分布或并不准确，之前的分布记录或是由于对其它蛇种误鉴所导致。考虑到赤峰锦蛇和棕黑锦蛇在分类上仍有诸多不确定性，本书将提供两者的分布图。绿色被认为是赤峰锦蛇的分布区域，蓝色则被认为是棕黑锦蛇的分布区域。多年来，韩国的种群被认为是两者的过渡类型（自然杂交个体）。最近的研究表明：基于目前的“线粒体DNA”和“微卫星”数据，韩国的种群被证实为棕黑锦蛇。

Specimen of unknown provenance (Matthew Most).

Habits and habitat: This species is diurnal and highly arboreal. Large adult ratsnakes have few enemies other than humans, with their automobiles and habitat destruction. Still, juveniles and small adults are known prey to various predators, from other snakes (such as *Elaphe carinata*), to hawks, owls, foxes, and swine. This species is found in a variety of habitats but is generally found in forested areas. Temperament varies from individual to individual. Some specimens, even wild ones, are docile, whereas other specimens may never become tame. This snake either "freezes" in place when approached or tries to crawl away. When prevented from fleeing, it may vibrate its tail, coil, and strike, and if handled, it may deliver a strong musk. Specimens will not hesitate to bite if continued to be provoked.

习性与栖息地：白昼活动，高度树栖。除人类外，大型锦蛇成年后几乎很少有天敌。栖息地遭受破坏，公路上来往的车辆是其生存面临的主要威胁。小亚成体和刚成年个体或成为其它蛇种（如：王锦蛇）、鹰、猫头鹰、狐狸和猪的猎食对象。赤峰锦蛇的生存环境较为多样化，主要以森林为主。不同个体间性情差异较大。部分野生个体也较为温顺，但也有个别极难驯服。人靠近时，本种通常于原地不动或试图逃走。若无法逃脱，其或会抖动尾部，盘曲身体并发动攻击。若被抓取，或分泌恶臭液体。如被反复滋扰，常毫不犹豫扑咬。

Diet and feeding habits: This species primarily feeds on rodents, birds, and their eggs; juveniles will also take lizards.

食性与喂养：主要以啮齿类、鸟类及鸟卵为食；小亚成体也会捕食蜥蜴。

E. anomala habitat in Beijing.

Specimen from Nanjing, Jiangsu, approximately the southern extent of the species' range.

Captive care: See general care at the beginning of this genus. In captivity, this species has been recorded to have a life span of 15-20 years.

人工饲养：具体情况请参阅本属介绍章节中的“人工饲养概述”版块。据记载，人工饲养下本种的寿命为15～20年。

Reproduction: Oviparous. Sexual maturity is achieved at 3-4 years. In captivity, specimens are provided with a brumation period of 10-15℃ for three months. Breeding is observed in the spring following brumation in late March to early April. The gestation period for *E. anomala* is 40-60 days, with eggs being laid in June and July. Clutch size ranges from 8-15 eggs with incubation of eggs at 23-25℃ for 40-60 days. Hatchlings measure about 280-450 mm in total length. This species has only been recorded and observed to lay a single clutch a year.

繁殖：卵生。于3～4年达到性成熟。人工饲养下，个体应在10～15℃下经历三个月的冬化期。据观察，交配在冬化后的来年3月下旬至4月上旬进行。母蛇怀卵40～60天。6～7月产卵。一窝卵8～15枚。孵化温度23～25℃，孵化期40～60天。刚出壳的幼蛇体长280～450毫米。据记载和观察，本种一年只产卵一窝。

Taxonomy and nomenclature: The distinction between this species and the Russian ratsnake (*E. schrenkii*) is in constant debate. Some authorities consider *E. anomala* as a subspecies of *E. schrenkii*. On the Korean peninsula, the taxonomy of this complex is even more chaotic. Some authorities think the entire population is an intergrade / hybrid composition (depending on whether the two species/ subspecies are distinct). However, recent studies have demonstrated that based on mtDNA and microsatellite data, the South Korean population aligns with the *E. schrenkii*.

分类与命名：关于本种与棕黑锦蛇的区别，争论始终存在。有权威人士认为赤峰锦蛇为棕黑锦蛇的亚种。在朝鲜半岛，情况则更为复杂：部分专家认为朝鲜半岛上的种群皆为两者的过渡类型（自然杂交个体）。然而，最近的研究表明：基于目前的"线粒体DNA"和"微卫星"数据，韩国的种群被证实为棕黑锦蛇。

A hybrid specimen of *E. anomala* x *E. schrenkii* from northern China (W. Scott Lupien).

Left: A juvenile *E. anomala* of unknown provenance (Richard Bartlett). Right: A juvenile *E. schrenkii* of unknown provenance (Matthew Most).

Juveniles of *E. schrenkii* and *E. anomala* are nearly impossible to distinguish. It is only with age that the distinctions become obvious. In captivity, several breeders have integrated these species.

棕黑锦蛇和赤峰锦蛇的小亚成体之间几乎无法区分。只有随着年龄的增长，两者间的差异才逐渐显现。人工饲养下，已有饲养者实现两者的杂交繁殖。

Conservation Status: The IUCN has not assessed this species.

保护状况：本种在《濒危物种红色名录》中被定为“未予评估”等级。

Comments: Though the name Korean ratsnake is often assigned to this species, it has been suggested that the Korean population may only consist of the Russian ratsnake, in which case, this common name is a bit misleading.

注释：尽管本种一大常见英文名为“Korean ratsnake（朝鲜锦蛇）”，但考虑到或许仅有棕黑锦蛇在朝鲜半岛有所分布，故该名称存在一定的误导性。

Scientific name / 学名: *Elaphe bimaculata* (Schmidt, 1925)

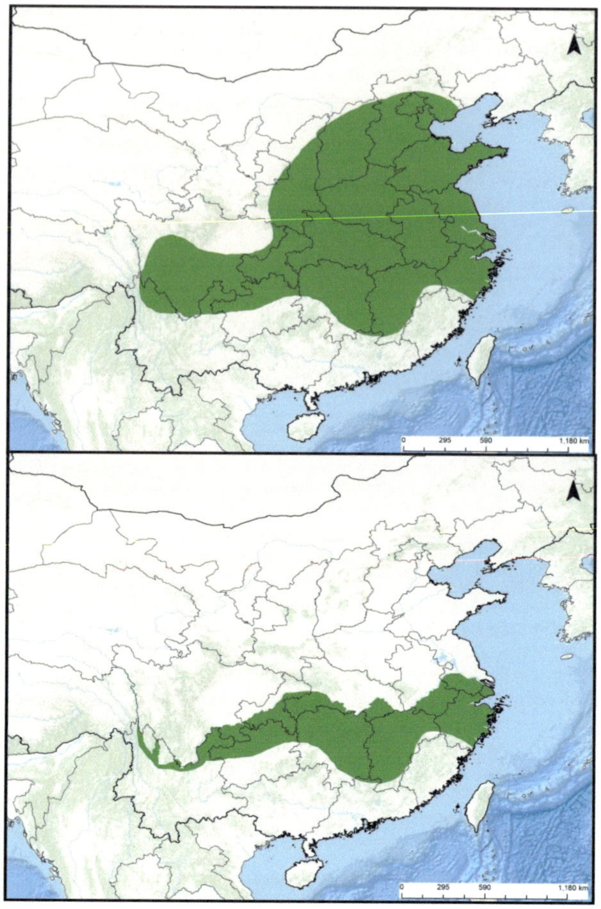

English name / 英文名: Twin-spotted ratsnake, Chinese Leopard Snake

Chinese name / 中文名: 双斑锦蛇 / shuāng bān jǐn shé

Etymology / 种名词源: *bimaculata* is a direct Latin translation meaning "two spots," referring to the twin-spotted pattern typically found on most specimens / "*bimaculata*"，拉丁语，意为"双斑"。该种名意指本种体背成对的斑纹。

Terra typica / 模式产地: Ningguo, Anhui, China (N30.634° E118.983°) / 中国，安徽省，宁国市

Length / 体长: Maturity in males is reached at 550 mm, females 750 mm; hatchlings emerge around 220-300 mm. Maximum lengths for males: 773 mm, 130 mm tail; and females 920 mm, 160 mm tail / 雄性体长550毫米时达到性成熟，雌性体长750毫米时达到性成熟。刚孵化的幼蛇体长220～300毫米。已知最大雄性个体全长773毫米，其中尾长130毫米；已知最大雌性个体全长920毫米，其中尾长160毫米。

Description: The cephalic scales are as follows: supralabials 8, 4th and 5th entering orbit; infralabials 10; preoculars 2; postoculars 2; temporals 2+3; scales smooth or feebly keeled posteriorly, in dorsal scale rows of 23:23:21. The average scale counts are ventrals 188-207, paired subcaudals 67-73, and a divided anal plate. On a tan to the yellowish background are two series of angular spots (some 40 in number) in reddish-brown outlined in black. Often the inner angles of the spots extend across the back to connect with the spot on the other side, forming a butterfly-like or bowtie-like pattern. The spots fuse from anterior to posterior in many individuals and produce well-defined reddish-brown stripes outlined with black. The lateral spots are not as well marked as the dorsal spots, but they also tend to fuse into stripes. Striping often occurs toward the tail. The belly is usually yellowish, with limited black spotting. Juveniles resemble adults.

(Rainer Fesser)

外观描述：鳞被：上唇鳞8（3-2-3式）；下唇鳞10；眶前鳞2；眶后鳞2；颞鳞2+3；背鳞23-23-21行，鳞片平滑或体后段微弱起棱；腹鳞通常188～207；尾下鳞67～73对；肛鳞二分。体背棕褐或淡黄色，正背具两行红褐色粗圆斑。圆斑镶有黑边，常左右相连呈哑铃状。许多个体的哑铃状斑块有纵向相连的趋势，形成镶有黑边的红褐色纵纹。体两侧也有斑块，不过不及正背斑块明显，同样有纵向相连的趋势。各斑块至尾部时通常已汇为纵纹。腹面通常为淡黄色，散有若干黑色斑点。小亚成体与成体在外观上相似。

Distribution: Precise species limits are unclear due to the ease of misidentifying this species with the very similar-looking *Elaphe dione*. Zhao (2006) lists the following for *E. bimaculata:* Anhui, Chongqing, Gansu, Hubei, Hunan, Jiangsu, Jiangxi, Shanghai, Sichuan, and Zhejiang. More recently, it is believed that those areas restricted to the south of the Yangtze River are most likely accurate. Previous records from Beijing, Gansu, Hebei, Henan, Inner Mongolia, Shanxi, Shaanxi, Shandong, and northern Sichuan are probably due to misidentification and representative of other species (e.g., *Elaphe dione*). A further complication in identification could come from hybridization between these two species, which could be observed in the area of their sympatry. The map on the top is the traditional *E. bimaculata* distribution recorded in Ji and Wen (2001); the map on the bottom is this same distribution but restricting the species to the south of the Yangtze River.

地理分布：由于在外观上与白条锦蛇较为相似，这对准确获悉双斑锦蛇于中国的分布造成影响。赵尔宓（2006）列出的分布区域包括：安徽、重庆、甘肃、湖北、湖南、江苏、江西、上海、四川和浙江。最近，本种于长江以南的分布被认为很可能是准确的；但先前于北京、甘肃、河北、河南、内蒙古、山西、陕西、山东和四川北部的分布记录则很可能是错误的，或是由于对其它蛇种（如白条锦蛇）误鉴所导致。在双斑锦蛇和白条锦蛇皆有分布的地理区域，两者的自然杂交个体更为辨识增加了难度。分布图上方的地图被认为是双斑锦蛇传统的分布区域（季达明&温世生，2001）；下方的地图则将双斑锦蛇的分布限于长江以南。

Typical habitat of *E. bimaculata* in Zhejiang.

Habits and habitat: Inhabits the relatively elevated region of the lower Yangtze. Daily activity is mostly nocturnal, but some individuals may occasionally be found in the morning or late afternoon, and daytime activity increases during the cooler months. This species is very docile by nature, and even wild specimens rarely bite when picked up.

习性与栖息地：栖息于长江下游海拔相对较高的区域。夜间活动为主，个别或偶尔晨昏活动。在气温较低的月份中，白天活动的时间明显增多。本种生性温顺，即使野生个体被抓取也很少扑咬。

Diet and feeding habits: Primarily feeds on rodents with a preference for smaller prey items, birds and their eggs; juveniles will also feed on lizards and their eggs.

食性与喂养：主要以啮齿类为食，偏爱鸟类、鸟卵及体型较小的猎物。小亚成体也捕食蜥蜴及其卵。

Captive care: See the general captive care for this species at the introduction of the genus. In captivity, this species has been recorded to have a life span of 10-15 years.

人工饲养：具体情况请参阅本属介绍章节中的"人工饲养概述"版块。据记载，人工饲养下本种的寿命为10～15年。

Reproduction: Oviparous. Sexual maturity is achieved at 2-3 years. In captivity, specimens are provided with a brumation period of 10-15°C for 3-4 months. Breeding is observed in the spring shortly after the following brumation. The gestation period for *E. bimaculata* is 48-55 days, with eggs being laid in June and July. Clutch size ranges from 2-12 eggs with incubation of eggs at 23-25°C for 25-40 days. This species has only been recorded and observed to lay a single clutch a year—triploid specimen reported by Rovatsos *et al.* 2018.

繁殖：卵生。于2～3年达到性成熟。人工饲养下，个体应在10～15°C下经历3～4个月的冬化期。据观察，交配在冬化后的春季进行。母蛇怀卵48～55天。6～7月产卵。一窝卵2～12枚。孵化温度23～25°C，孵化期25～40天。据记载和观察，本种一年只产卵一窝。Rovatsos等人（2018）曾报道三倍体双斑锦蛇。

Taxonomy and nomenclature: Based on morphological features, *E. bimaculata* and *E. dione* were considered conspecific by Brunner (1995). Schmidt (2000) confirmed the species status of *bimaculata* due to differences in hemipenis morphologies.

分类与命名：根据形态特征，Brunner（1995）曾认为双斑锦蛇和白条锦蛇为同一物种。由于两者半阴茎形态上的差异，Schmidt（2000）证实双斑锦蛇为有效种。

Elaphe bimaculata of unknown provenance (Matthew Most).

Similar species: Can easily be confused with *E. dione*, but Schulz (1986) gives several distinctive characteristics to use when identifying: 1) *E. bimaculata* has 8-10, seldomly 11 infralabial scales: *E. dione* normally has 11-12, seldomly 13. 2) The markings of the head of *E. bimaculata* merge into two small posterior stripes in the neck region. In *E. dione,* these stripes are broader and shorter. 3) *E. bimaculata* has 18-20 teeth in the upper jaw, *E. dione* only 15-17. 4) The hemipenes of both species are different, but Schulz does not mention how they differ. 5) The spots on *E. bimaculata* have black borders, whereas these spots on *E. dione* are not bordered (open). *E. bimaculata* has two primary pattern morphs, dumbbell and twin spotted.

相似物种：极易与白条锦蛇相混淆，Schulz（1986）提供了数个显著特征用以鉴别：1）双斑锦蛇下唇鳞通常为8～10枚，少数11枚；白条锦蛇下唇鳞通常为11～12枚，少数13枚。2）双斑锦蛇头部的斑纹可连接颈后部深处的斑点，形成两条纵纹；相较之丁，白条锦蛇此处的斑纹更宽、更短。3）双斑锦蛇有18～20颗上颌齿，而白条锦蛇只有15～17颗。4）两者的半阴茎不同，但Schulz并未对此详细说明。5）双斑锦蛇身上的斑点镶有黑边，白条锦蛇身上的斑点则无黑边。双斑锦蛇身上的斑纹主要为两种形态：“哑铃状”或“双斑点”。

Conservation Status: This species is considered Least Concern (LC) by the IUCN.

保护状况：本种在《濒危物种红色名录》中被定为“无危”等级。

Above: *E. bimaculata* specimen (Rainer Fesser)

Below: *E. dione* specimens from Beijing (W. Scott Lupien)

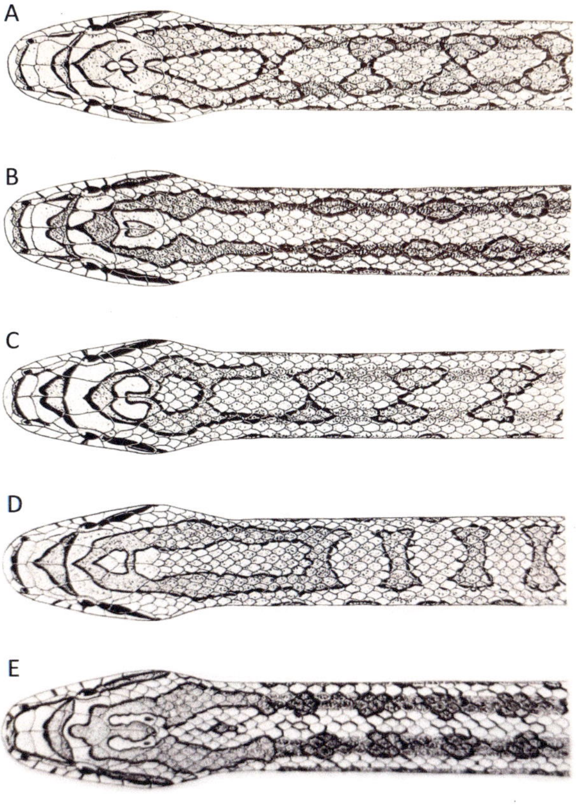

From Schulz, 1996:

A - C: Examples of *E. bimaculata* with relatively clear longitudinal dorsal stripes.

D: Spotted phase of *E. bimaculata.*

E. *E. dione* from Anhui.

Scientific name / 学名: *Elaphe carinata* (Gunther, 1864)

English name / 英文名: King ratsnake, Stinking Goddess, Keeled ratsnake

Chinese name / 中文名: 王锦蛇 / wáng jǐn shé

Etymology / 种名词源: *carinata* is Latin for "keeled," referring to the heavily keeled scales of this species / "*carinata*"，拉丁语，意为"有棱的"。该种名意指王锦蛇强烈起棱的鳞片。

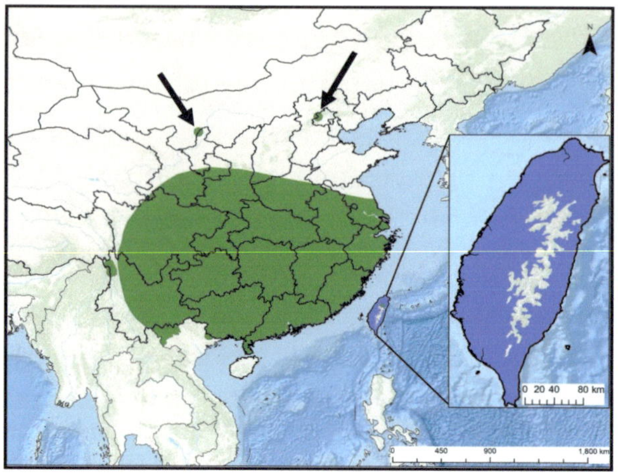

green = *E. c. carinata*, blue = *E. c. yonaguniensis*

Terra typica / 模式产地: Lu Shan, Jiujiang, Jiangxi, China (N29.5° E116.0°)/ 中国，江西省，九江市，庐山

Length / 体长: Maturity at 1400-1700 mm; Hatchlings at 300-390 mm. Maximum 2440 mm / 体长1400～1700毫米时达到性成熟。刚孵化的幼蛇体长300～390毫米。已知最大个体体长2440毫米。

Description: The cephalic scales are as follows: supralabials 8, rarely 7; preoculars 2, rarely 3; postoculars 2, rarely 3; anterior temporals 2, occasionally 3; posterior temporals 3, occasionally 2; scales heavily keeled in an average dorsal scale row arrangement of 23:23:19. There are 208-223 ventrals, 80-97 paired subcaudals, and a divided anal plate. Ontogenetic change from juvenile to adult. Adults are brownish-yellow, olive-yellow, or bronze in color, with heavily keeled scales bordered in black, giving a spotted or sometimes partially banded appearance. Melanistic specimens frequently occur in normally colored populations. A fairly unique characteristic of this species, which no other members of *Elaphe* are known to exhibit, is that this species may have an elliptical pupil when the pupil is constricted (round when dilated). Juveniles are much

Left: Specimen flattening its head in a triangular fashion, in order to mimic a venomous species (Dan Rosenberg). Right: Close up of the eye, showing the elliptical pupil that this species sometimes exhibits in bright conditions (Guillaume Blanchet).

paler and more uniform in coloration, usually a shade of tan with a few scattered black dots, which on the anterior half of the body, form short and narrow cross lines.

外观描述：鳞被：上唇鳞8，少数7枚；眶前鳞2，少数3；眶后鳞2，少数3；前颞鳞2，偶有3；后颞鳞3，偶有2；背鳞通常23-23-19行，强烈起棱；腹鳞208～223；尾下鳞80～97对；肛鳞二分。小亚成体与成体在外观上有较大差异。成体背面黄褐、橄榄黄或古铜色，鳞片强烈起棱且镶有黑边，形成点状或带状的黑色斑块。相较于成体，小亚成体体色较暗淡，通身颜色较单调，常为棕褐色并散有黑色斑点，于体前段形成细短横斑。本种常可见体色变异个体，如黑化个体。在锦蛇属所有蛇种中，王锦蛇有一独树一帜的特征：其瞳孔在收缩时为直立椭圆形（扩张时为圆形）。

Distribution: This is a very widespread species. Throughout Greater China, it is found in Anhui, Beijing, Chongqing, Fujian, Gansu, Guangdong, Guangxi, Guizhou, Hebei, Henan, Hubei, Hunan, Jiangsu, Jiangxi, Ningxia, Shaanxi, Shanxi, Shanghai, Sichuan, Shandong, Taiwan, Tianjin, Yunnan, and Zhejiang, but curiously absent from Hong Kong and Hainan Island. There is a disjunct population found in the vicinity north and west of Beijing. Outside of China, it is found in northern Vietnam and Japan. Found from 0 - 3200 m.

地理分布：广布种。在中国，分布于安徽、北京、重庆、福建、甘肃、广东、广西、贵州、河北、河南、湖北、湖南、江苏、江西、宁夏、陕西、山西、上海、四川、山东、台湾、天津、云南和浙江。海南和香港未有分布记录，这点颇令人意外。在北京西、北周边区域还有间断分布。中国以外，分布于越南北部和日本。垂直分布0～3200米。

Habits and habitat: Primarily terrestrial, inhabiting montane forests, bamboo thickets, rocky valleys, highland agricultural areas, meadows, and may be found near human dwellings. Usually inhabiting thickly forested areas of high mountainsides. Nocturnal and diurnal. When not actively foraging, they frequently seek refuge in mammal burrows or rock piles. This species is quite a fighter. Upon immediate discovery, specimens will attempt to flee quickly, but if escape is not possible, they will hiss, puff up with air, strike repeatedly, and readily musk when cornered. If handled, individuals will bite, chew, and musk in attempts to evade prolonged capture.

习性与栖息地：地栖性为主。栖息于山林、竹林、岩壑、农耕高地、草甸，甚至人类住所附近；其中，最典型的栖息地貌为高山密林。昼夜皆会活动。不觅食时，常躲藏在哺乳动物所挖掘的洞穴或岩石堆中。生性好斗。如被发现，会立即设法逃离。如无法逃脱，则会鼓起身体、张嘴发出嘶嘶声并反复发动攻击。若被抓取，会扑咬并分泌恶臭液体。

Many specimens will gape, hiss, and strike repeatedly when they feel threatened (Dan Rosenberg).

Diet and feeding habits: Opportunistic feeder, preying on nearly anything it can subdue. Although rodents, birds, and bird eggs are common prey items, king ratsnakes are also ophiophagus, consuming both harmless and venomous species. Kills prey via constriction if necessary. Pope (1935) also recorded an individual with a Megophrys frog in its stomach.

食性与喂养：机会主义者，捕食几乎所有能制服的猎物。常见猎物包括啮齿类、鸟类和鸟卵。有时也猎食同类，无毒和有毒蛇种都可能成为王锦蛇的食物。必要时会绞杀猎物。Pope（1935）曾记载在王锦蛇的胃中发现有未消化完全的角蟾属（*Megophrys*）物种。

Captive care: See the general care section at the introduction of this genus. Do not keep this species in a communal setup. Specimens will readily eat cagemates. In captivity, this species has been recorded to have a life span of 15-20 years.

人工饲养：具体情况请参阅本属介绍章节中的“人工饲养概述”版块。考虑到王锦蛇会吞食同类，故切勿群养。据记载，人工饲养下本种的寿命为15～20年。

E. carinata consuming a sharp-nosed viper (*Deinagkistrodon acutus*) in Taiwan (Pingtung Forest District Office, Forestry Bureau 林務局屏東林管處).

A juvenile *E. carinata* from Guangdong (Konrad Mebert).

Reproduction: Oviparous. Sexual maturity is achieved at 4-5 years. In captivity, specimens are provided with a brumation period of 10-15℃ for three months. Breeding is observed in the spring shortly after the following brumation. The gestation period for *E. carinata* is 60-70 days, with eggs being laid in June and July. Clutch size ranges from 5-20 eggs with incubation of eggs at temperatures of 23-25℃ for 60-80 days. In the wild, this species will seek out decaying vegetation to lay their eggs. This species has only been recorded and observed to lay a single clutch a year. The species experience ontogenetic changes. The 30-40 cm young are colored and patterned differently from adults, being pale with indistinct markings and light dorsolateral stripes.

繁殖：卵生。于4～5年达到性成熟。人工饲养下，个体应在10～15℃下经历3个月的冬化期。据观察，交配在冬化后的春季进行。母蛇怀卵60～70天。6～7月产卵。一窝卵5～20枚。孵化温度23～25℃，孵化期60～80天。自然环境中，王锦蛇会寻找腐烂的植被来产卵。据记载和观察，本种一年只产卵一窝。30～40厘米体长的小亚成体与成体在体色和斑纹上有较大不同。小亚成体体色较暗淡，体背具浅色侧纵线并散有不明显斑纹。

Taxonomy and nomenclature: There are two recognized subspecies. *Elaphe c. souliei* conforms with *E. c. deqenensis* in the locality and morphological data. For this reason, *E. c. deqenensis* is synonymous with *E. c. souliei.* However, neither of these subspecies are valid as Guo et al. (2012) synonymized *E. c. deqenensis* with *E. c. carinata.*

E. c. carinata : This subspecies is found throughout the vast majority of China, including Beijing, south to northern Vietnam.

E. c. deqenensis / souliei (Mocquard, 1897); **No longer recognized. Type locality: Tsekou (Yanmen, between Bingzhongluo and Deqen, Hengduan Shan Mountains, Yunnan, China. Only known from the type locality and adjacent areas.

E. c. yonaguniensis (Takara, 1962); Type locality: Yonaguijima Island, Ryukyu Archipelago, Japan. This species is found in Japan (Yonagunijima, Ryukyu Archipelago) and Taiwan.

分类与命名：存在两个公认的亚种。*Elaphe c. souliei* 与 *E. c. deqenensis* 在分布与形态学数据方面一致，故 *E. c. deqenensis* 实为 *E. c. souliei* 的同物异名。之后的研究表明：*E. c. deqenensis* 又为 *E. c. carinata* 的同物异名（郭鹏等人，2012），故 *E. c. souliei* 与 *E. c. deqenensis* 皆不再为有效亚种。

E. c. carinata: 本亚种在中国绝大多数地区（包括北京）有分布，向南的分布可一直延伸至越南北部。

E. c. deqenensis / souliei (Mocquard, 1897)：**不再为公认的亚种。模式产地：燕门乡（地处云南省迪庆藏族自治州德钦县南部）。已知仅分布于模式产地及其周边区域。

E. c. yonaguniensis (Takara, 1962)：模式产地：琉球群岛中的与那国岛。除与那国岛外，本亚种在中国台湾也有分布。

Comments: The presence of exceptionally large anal glands and the rapid discharge of their noxious contents when threatened is how the species earns one of its common names, the Stinking Goddess. In Taiwan, this species is known by a different Chinese name, it is known as: 臭青公 chòu qīng gōng.

注释：无论中外，王锦蛇的各式俗名都常与“臭”联系在一起，这主要同本种巨大的肛腺和受威胁时能快速分泌恶臭液体有关。在中国台湾，本种被称为“臭青公”。

Conservation Status: The IUCN has not assessed this species.

保护状况：本种在《濒危物种红色名录》中被定为“未予评估”等级。

Top left: Specimen from western Hubei. Top right: Specimen from northern Fujian.
Bottom: Specimen from northern Taiwan.

Right: Specimen from northern Taiwan (Kevin Caldwell).

Below: Specimen from Chengdu, Sichuan (Gernot Vogel).

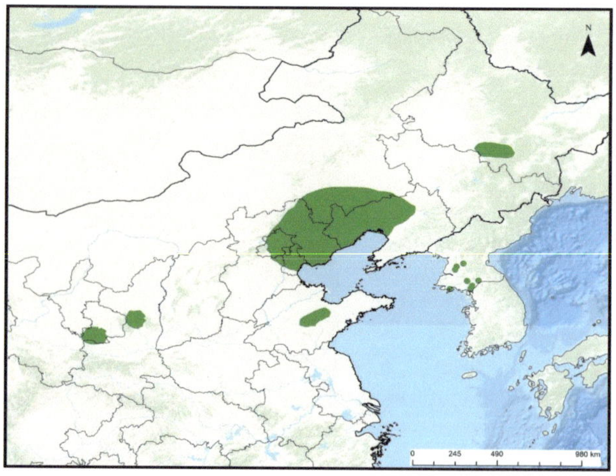

Scientific name / 学名: *Elaphe davidi* (Sauvage, 1884)

English name / 英文名: David's ratsnake

Chinese name / 中文名:团花锦蛇 / tuán huā jǐn shé

Etymology / 种名词源: The specific name, *davidi* is named in honor of French naturalist Père Armand David (1826-1900), who collected the first specimen, as well as numerous other species in China, including the Giant Panda, Chinese Giant Salamander, and Père David's Deer / 种名"*davidi*"是为了纪念法国博物学家阿尔芒•戴维德（Père Armand David, 1826-1900），他是团花锦蛇首枚标本的采集者。同时，他还是许多中国特有物种如大熊猫、中国大鲵、麋鹿的发现者。

Terra typica / 模式产地: "China" / 中国

Length / 体长: Maturity in males is reached at 550 mm and females 800 mm; Hatchlings emerge around 250-340 mm. Maximum sizes recorded for males are 1410 mm (183mm tail), and females 950 mm (156 mm tail) / 雄性体长550毫米时达到性成熟，雌性体长800毫米时达到性成熟。刚孵化的幼蛇体长250～340毫米。已知最大雄性个体全长1410毫米，其中尾长183毫米；已知最大雌性个体全长950毫米，其中尾长156毫米。

Description: The cephalic scales are as follows: supralabials 8, 4th and 5th entering orbit; preoculars 2; postoculars 2, occasionally 3; anterior temporals 2 or 3, rarely 1; scales heavily keeled in dorsal scale rows of 23: 23 or 25: 19 (rarely 21 or 17); ventrals 171-173 (males), 173-181 (females); subcaudals 65-69 (males), 58-64 (females); divided anal plate; subcaudals paired. The dorsal scales are strongly keeled, perhaps indicating a relationship to *E. carinata*. The background color of the body is yellowish to tan, and there is a series of dark brown dorsal blotches edged with black. Smaller lateral blotches are present. A dark band arches across the head in front of the eyes and continues behind the eyes to the angles of the jaws. Other dark head markings in complex patterns are found in young snakes. The belly is mottled with light gray, and every second or third scale has a dark spot at the end. Juveniles resemble adults.

外观描述：鳞被：上唇鳞8（3-2-3式）；眶前鳞2；眶后鳞2，偶有3；前颞鳞2或3，少数1；背鳞23-23（25）-19（少数17或21）行，强烈起棱；雄性腹鳞171～173，雌性腹鳞173～181；雄性尾下鳞65～69对，雌性尾下鳞58～64对；肛鳞二分。背鳞强棱表明团花锦蛇或同王锦蛇有较近的亲缘关系。体背淡黄或棕褐色，有三行深褐色镶有黑边的椭圆形斑块。正背一行的斑块大于体两侧的斑块。头背有一深色弧形斑纹。该斑纹入眼并延伸直达口角。年轻个体头部的斑纹更为复杂。腹面浅灰色，散有深色斑点。小亚成体与成体在外观上相似。

Distribution: Within Greater China, found in Beijing, Hebei, Heilongjiang, Inner Mongolia, Jilin, Liaoning, Shaanxi, Shandong and Tianjin. Records from Shanxi are expected to be in error. Outside of China, it has been reported in North Korea. Localities near the southeastern border of North Korea have been viewed with skepticism as many herpetologists think the species would have been found in South Korea if these records are valid.

地理分布：在中国，分布于北京、河北、黑龙江、内蒙古、吉林、辽宁、陕西、山东和天津。本种于山西省的分布记录或并不正确。中国以外，朝鲜据报道也有本种分布，不过两栖和爬行动物学家对团花锦蛇于朝鲜东南部的分布持怀疑态度。毕竟若确有分布，那韩国也将大概率有本种的分布记录。

Habits and habitats: Diurnal. Found in fields, mountains, open river valleys, orchards, grassy fields, and rocky areas. This species is docile by nature, and even wild specimens rarely bite when picked up.

习性与栖息地：白昼活动。可见于田间、山区、开阔河谷、果园、草地和多岩石区域。生性温顺，即使野生个体被抓取也很少扑咬。

The dry, rocky habitat north and west of Beijing, especially those areas with dilapidated sections of the Great Wall make for excellent habitat for *E. davidi*.

Diet and feeding habits: This species is an opportunistic feeder that raids both bird and rodent nests. *E. davidi* will feed on rodents, birds, and their eggs. Birds and bird eggs are preferred during the summer months.

E. davidi has several morphological adaptations similar to other egg-eating species, including the shape of the hypapophyses penetrating the esophagus, the proportions and kinetics of some skull elements, the number of anterior dorsal scale rows, modifications of the esophageal lining, and a cranial shift of the heart and the anterior liver tip (Helfenberger and Schatti, 1998).

食性与喂养：机会主义者，以鸟类、鸟卵和啮齿类为食。夏季时尤喜食鸟类以及鸟卵。在脊椎（椎体下突）、头骨、心脏、肝脏、食管内膜、体前段背鳞行数等方面的形态特征与食卵蛇种十分相似（Helfenberger和Schatti，1998）。

Captive care: See the general care section at the introduction of this genus. In captivity, this species has been recorded to have a life span of 10-15 years.

人工饲养：具体情况请参阅本属介绍章节中的“人工饲养概述”版块。据记载，人工饲养下本种的寿命为10~15年。

Reproduction: Oviparous. Sexual maturity is achieved at 3-4 years. In captivity, specimens are provided with a brumation period of 10-15° C for three months. Breeding is observed in the spring shortly after the following brumation. The gestation period for *E. davidi* is 60 days, with eggs being laid in June and July.

A young *E. davidi* raiding a bird nest on 5 July.

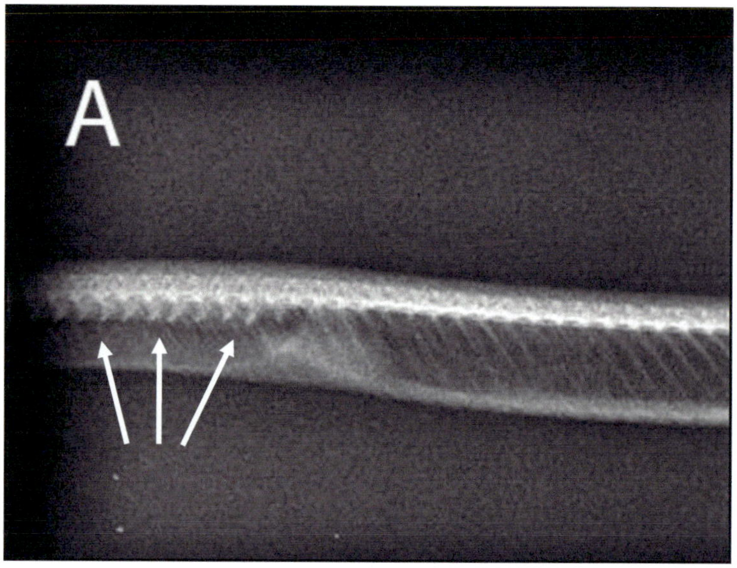

Radiograph of an *E. davidi* (A = anterior) illustrating the hypapophyses on the ventral side of the anterior portion of the spine (Matthew Most).

An adult *E. davidi* constricting a hamster (W. Scott Lupien).

Clutch size ranges from 5-8 eggs with incubation of eggs at 23-25°C for 60-70 days. This species has only been recorded and observed to lay a single clutch a year. Juveniles resemble miniature adults in the pattern.

Female *E. davidi* with eggs (above; Matthew Most).

繁殖：卵生。于3～4年达到性成熟。人工饲养下，个体应在10～15℃下经历3个月的冬化期。据观察，交配在冬化后的春季进行。母蛇怀卵60天左右。6～7月产卵。一窝卵5～8枚。孵化温度23～25℃，孵化期60～70天。据记载和观察，本种一年只产卵一窝。小亚成体身上的斑纹同刚成年个体相似。

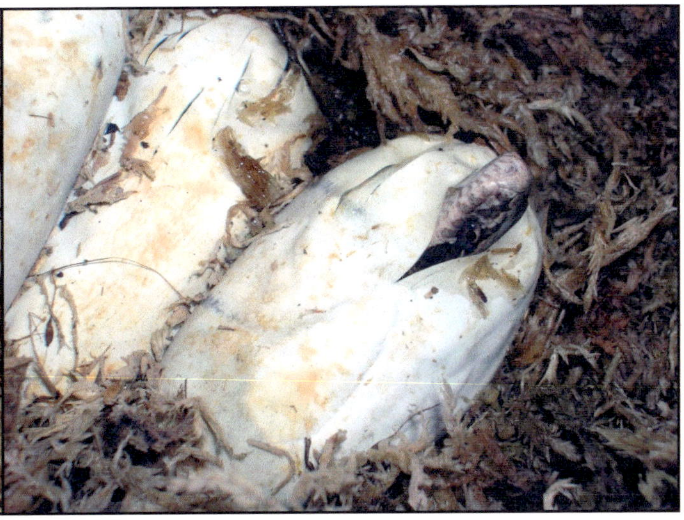

Juvenile *E. davidi* hatching from eggs. Beijing, China (W. Scott Lupien).

Taxonomy and nomenclature: There are no known subspecies.

分类与命名： 目前无已知的亚种。

Similar species: This species could be easily confused with the genus *Gloydius* based on coloration and the presence of rough scales. This may serve as an ecological advantage in the success of this species within its natural range. Globally, many colubrids are known to have a similar coloration of local viper species, such as the harmless colubrids, *Pantherophis emoryi,* and *Lampropeltis calligaster,* in the Midwest of the United States of America, where both species are sympatric with the venomous *Sistrurus catenatus.* Dull colored variations of *E. dione,* or individuals that have emerged from underground and still covered with mud or dirt, giving off a "rough" appearance, could also be confused for this species by an untrained eye. Additionally, if only a portion of the animal can be seen, positive identification is further complicated. As a result, caution should be exercised if one is in an area where both *E. davidi* and *G. brevicaudus* can be found. As for *E. dione,* they are easily distinguished by having smooth scales versus rough scales (present in *E. davidi*).

Upper images: *E. davidi* (left; W. Scott Lupien), *E. dione* (right; W. Scott Lupien)
Lower images: *Gloydius brevicaudus*

Above: A young *E. davidi* specimen (Matthew Most); Below: Specimen from Beijing (W. Scott Lupien).

相似物种：极易与亚洲蝮属（*Gloydius*）的蛇种相混淆，这甚至被认为是某种生态优势。全球范围内，很多游蛇在体色上都与同一生态环境中的某些蝰蛇较为相似。比如美国中西部两种无毒的游蛇——大平原鼠蛇（*Pantherophis emoryi*）和草原王蛇（*Lampropeltis calligaster*）就与当地的有毒蝰蛇——北美侏儒响尾蛇（*Sistrurus catenatus*）在外观上相似。如果仅身体某一部位可见，辨识则更为困难。故在短尾蝮（*Gloydius brevicaudus*）和团花锦蛇皆有分布的地域活动时应格外小心。体色黯淡的白条锦蛇，或身上裹有污泥的白条锦蛇（造成背鳞强烈起棱的假象）也或被非专业人士误认为团花锦蛇。其实，白条锦蛇背鳞平滑，较易与团花锦蛇（背鳞强烈起棱）进行区分。

Conservation Status: Not assessed by IUCN as of 2020.

保护状况：本种在《濒危物种红色名录》中被定为“未予评估”等级。

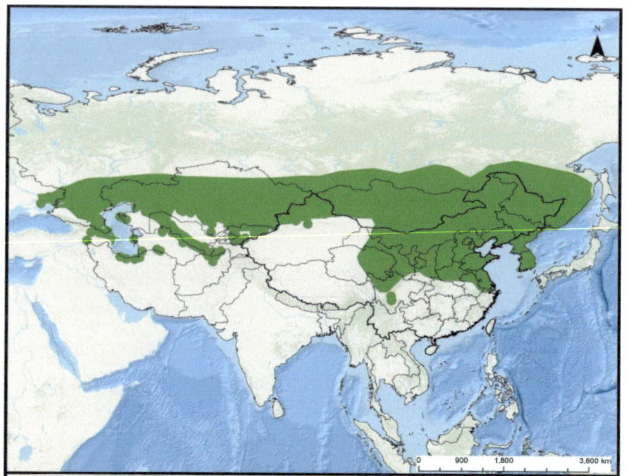

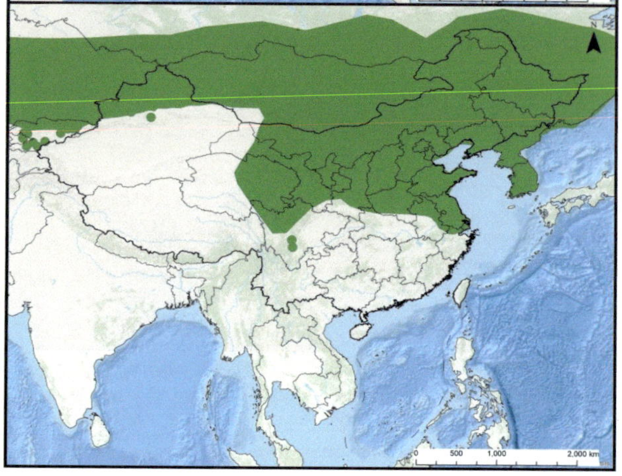

Scientific name / 学名: *Elaphe dione* (Pallas, 1773)

English name / 英文名: Dione's ratsnake, Steppes Ratsnake

Chinese name / 中文名称: 白条锦蛇 / bái tiáo jǐn shé

Etymology / 种名词源: Named after Dione, a Titan goddess in Greek mythology and the mother of the goddess Aphrodite / 以希腊神话中的女神狄俄涅（Dione）的名字来命名。狄俄涅为爱和美神阿佛洛狄忒（Aphrodite）之母。

Terra typica / 模式产地: According to Brunner (1995), the type locality can be restricted to the salt steppe of the Caspian Basin, east at the mouth of Ural River / 根据Brunner（1995），本种的模式产地限于乌拉尔河河口东部里海盆地的盐渍草原。

Length / 体长: Maturity at 400 mm (males) and 760 mm (females); Hatchlings emerge around 160-250 mm. The maximum lengths for males are 1410 mm (183 mm tail) and 1026 mm (150 mm tail) for females / 雄性体长400毫米时达到性成熟，雌性体长760毫米时达到性成熟。刚孵化的幼蛇体长160～250毫米。已知最大的雄性个体全长1410毫米，其中尾长183毫米；已知最大的雌性个体全长1026毫米，其中尾长150毫米。

Description: The cephalic scales are as follows: supralabials 8, rarely 9; preoculars 2, very rarely 3; postoculars 2; anterior temporals 2, rarely 3; posterior temporals 3 or 4; dorsal scale rows 25 (sometimes 23); smooth and shiny, or feebly keeled. Ventrals 171-214 (seldom over 204); paired subcaudals 51-78; anal plate divided, rarely single. Highly variable in coloration and pattern, the head markings are distinctive and resemble a dark horseshoe or rounded spearpoint that runs from the napes of the neck over the top of the head to the frontal scale. Another dark band runs from eye-to-eye and continues to the jaws. The body's pattern usually consists of

Specimens from Beijing.

some combination of four weakly marked dark stripes and narrow dark and light bands over the back. The markings may be distinct and edged with black or may fade into the background color. Other adults appear to be covered with a dark and light reticulated pattern. Juveniles resemble adults.

外观描述：鳞被：上唇鳞8，少数9；眶前鳞2，少数3；眶后鳞2；前颞鳞2，少数3；后颞鳞3或4；体中段背鳞25行（个别23行）；背鳞平滑或微弱起棱。腹鳞171～214（极少超过204）；尾下鳞51～78对；肛鳞二分，少数个体肛鳞完整。本种体色、斑纹极为多变。头背斑纹较独特：有一形似马蹄的深色斑块从枕部延伸至额鳞，另有一深色弧形斑纹入眼并在眶后延伸斜达口角。体背通常有四条不明显的深色纵纹并散有若干深浅相间的短横斑。短横斑或较明显并镶有黑边，或与底色相近较难分辨。某些成体背面具深浅相间的网状纹。小亚成体与成体在外观上相似。

Distribution: Precise species limits within Greater China are unclear due to the ease of misidentifying this species with the very similar-looking *Elaphe bimaculata*. In China, *E. dione* is most probably restricted to areas north of the Yangtze River. Records from Anhui, southern Jiangsu, and southern Sichuan are probably due to misidentification and representative of other species (e.g., *Elaphe bimaculata*). A further complication in identification could come from hybridization between these two species, which could be observed in the area of their sympatry (Simonov *et al.* 2018).

Example of habitat for *E. dione* in Mongolia.

However, according to Zhao (2006) and Zhao & Adler (1993), officially, the known distribution is Anhui, Beijing, Gansu, Hebei, Heilongjiang, Henan, Hubei, Inner Mongolia, Jiangsu, Jilin, Liaoning, Ningxia, Qinghai, Shaanxi, Shanxi, Shandong, Shanghai, Sichuan, Tianjin, Tibet, and Xinjiang. Outside of Greater China, it is found in Afghanistan, Azerbaijan, Georgia, Iran, Kazakhstan, the Korean peninsula, Kyrgyzstan, Mongolia, Russia, Tajikistan, Turkmenistan, Ukraine, and Uzbekistan. Found as high as 3520 m.

地理分布：由于在外观上与双斑锦蛇较为相似，这对准确获悉白条锦蛇于中国的分布造成影响。在中国，本种的分布很可能局限于长江以北地区。其在安徽、江苏南部、四川南部的分布记录或是由于对其它蛇种（如双斑锦蛇）误鉴所导致。在白条锦蛇和双斑锦蛇皆有分布的地理区域，两者的自然杂交个体也为辨识增加了难度（Simonov等人，2018）。根据赵尔宓（2006）和赵尔宓 & Kraig Adler（1993），在中国，本种已知分布于安徽、北京、甘肃、河北、黑龙江、河南、湖北、内蒙古、江苏、吉林、辽宁、宁夏、青海、陕西、山西、山东、上海、四川、天津、西藏和新疆。中国以外，本种分布于阿富汗、阿塞拜疆、格鲁吉亚、伊朗、哈萨克斯坦、朝鲜半岛、吉尔吉斯斯坦、蒙古、俄罗斯、塔吉克斯坦、土库曼斯坦、乌克兰和乌兹别克斯坦。垂直分布高达海拔3520米。

Habits and habitats: Primarily diurnal but may become nocturnal in the hot summer months. Adult *E. dione* seldom bites if disturbed—instead, they vigorously vibrate their tails and spray musk if handled.

习性与栖息地：白昼活动为主，夏季可能会夜出活动。成体遇滋扰也很少扑咬。若被抓取，会剧烈抖动尾部并分泌恶臭液体。

Typical habitat of *E. dione* from Mongolia.

Diet and feeding habits: Primarily feed on rodents with a preference for smaller prey items, birds, and eggs. Has also been noted to eat smaller snakes in captivity, with mates being consumed during breeding if left unattended.

食性与喂养：主要以啮齿类为食，偏爱鸟类、鸟卵及体型较小的猎物。人工饲养下，会吞食群养环境中体型较小的蛇。

Captive care: See the general care section at the introduction of this genus. In captivity, this species has been recorded to have a life span of 10-15 years.

人工饲养：具体情况请参阅本属介绍章节中的"人工饲养概述"版块。据记载，人工饲养下本种的寿命为10～15年。

Reproduction: Oviparous. Sexual maturity is achieved at 3-4 years. In captivity, specimens are provided with a brumation period of 10-15°C for three months. Breeding is observed in the spring shortly after the following brumation. The gestation period for *E. dione* is 58-89 days, with eggs being laid in July and August. Clutch size ranges from 6-21 eggs with incubation of eggs at 23-25°C for 15-38 days. This species has a much longer gestation period with a shorter incubation period with eggs turning clear before hatching. Females are also capable of producing 1-2 clutches a year.

繁殖：卵生。于3～4年达到性成熟。人工饲养下，个体应在10～15℃下经历3个月的冬化期。据观察，交配在冬化后的春季进行。母蛇怀卵58～89天。7～8月产卵。一窝卵6～21枚。孵化温度23～25℃，孵化期15～38天。白条锦蛇怀卵期较长但孵化期较短。幼蛇出壳前，卵壳呈透明状。本种一年产卵1～2窝。

Gravid female *E. dione* from the Beijing area (W. Scott Lupien).

Taxonomy and nomenclature: No subspecies are recognized.

分类与命名：目前无已知的亚种。

Similar species: Can easily be confused with *E. bimaculata*, but Schulz (1986) gives several distinctive characteristics to use when identifying: 1) *E. bimaculata* has 8-10, seldomly 11 lower labial scales: *E. dione* normally has 11-12, seldomly 11-13 lower labials. 2) The markings of the head of *E. bimaculata* merge into two small posterior stripes in the neck region. In *E. dione,* these stripes are broader and shorter. 3) *E. bimaculata* has 18-20 teeth in the upper jaw, *E. dione* only 15-17. 4) The hemipenes of both species are different, but Schulz does not mention how they differ. 5) The spots on *E. bimaculata* have black borders, whereas these spots on *E. dione* are not bordered (open). *E. bimaculata* has two primary pattern morphs, dumbbell and twin spotted. The species could also be confused with *Elaphe davidi* but is easily distinguished from this species by having smooth scales.

Specimen from Seoul, South Korea.

相似物种： 极易与双斑锦蛇相混淆。对此，Schulz（1986）提供了数个显著特征用以鉴别：1）双斑锦蛇下唇鳞通常为8～10枚，少数11枚；白条锦蛇下唇鳞通常为11～12枚，少数13枚。2）双斑锦蛇头部的斑纹可连接颈后部深处的斑点，形成两条纵纹；相较之下，白条锦蛇此处的斑纹更宽、更短。3）双斑锦蛇有18～20颗上颌齿，而白条锦蛇只有15～17颗。4）两者的半阴茎不同，但Schulz并未对此详细说明。5）双斑锦蛇身上的斑点镶有黑边，白条锦蛇身上的斑点则无黑边。双斑锦蛇身上的斑纹主要为两种形态："哑铃状"或"双斑点"。同时，白条锦蛇也或被误认为团花锦蛇。其实，白条锦蛇背鳞平滑，较易与团花锦蛇（背鳞强烈起棱）进行区分。

Conservation Status: The species is ranked as Least Concern (LC) by the IUCN.

保护状况： 本种在《濒危物种红色名录》中被定为"无危"等级。

Comments: Dione's ratsnake is the most widespread species of the genus *Elaphe*. The species is extremely variable. In the pet trade, several colors and pattern morphs have been produced.

注释： 白条锦蛇为锦蛇属下分布最广的蛇种。本种外观上极为多变。宠物市场已培育出诸多不同体色和斑纹的个体。

Melanistic *E. dione* specimens from the Altai mountain range at the intersection of Russia x Kazakhstan x Mongolia x China (Matthew Most).

Scientific name / 学名: *Elaphe schrenkii* (Strauch, 1873)

English name / 英文名: Russian ratsnake, Siberian ratsnake, Amur ratsnake

Chinese name / 中文名: 棕黑锦蛇 / zōng hēi jǐn shé

Etymology / 种名词源: Named after Dr. Leopold von Schrenk (1826-1894), the celebrated Russian naturalist and explorer of Amurland / 以俄罗斯著名博物学家和探险家Leopold von Schrenk（1826-1894）博士的名字来命名。

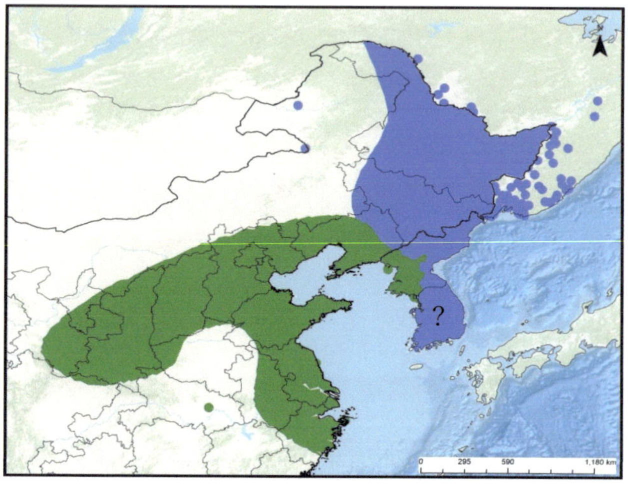

green = *E. anomala*, blue = *E. schrenkii*

Terra typica / 模式产地: Khinggan Military Post, Russia / 俄罗斯，兴安军邮站。

Length / 体长: Maturity at 1000-1100 mm; Hatchlings emerge around 310-420 mm; Maximum length of males 1600 mm (265 mm tail); females 1574 mm (224 mm tail) / 体长1000～1100毫米时达到性成熟。刚孵化的幼蛇体长310～420毫米。已知最大雄性个体全长1600毫米，其中尾长265毫米；已知最大雌性个体全长1574毫米，其中尾长224毫米。

Description: The dorsal scales are in 21-23 rows at midbody and are distinctly keeled towards the midline. There are 208-230 ventrals and 61-76 paired subcaudals; the anal plate is divided. There is a small pre-subocular wedged between the front of the eye and the supralabial. This species is a large, heavy-bodied snake with indistinct head, black and yellow (or white) striped lips, and a body pattern of big rounded black saddles separated by narrow yellow or white oblique bars, vaguely similar to a chain in effect. The saddles usually number between 20-30.

外观描述：大型蛇。鳞被：体中段背鳞21～23行，中央若干行起棱明显；腹鳞208～230；尾下鳞61～76对；肛鳞二分；具一枚小的眶前下鳞。头、颈区分不明显。唇部具黑黄或黑白条纹。体背具黑色圆鞍形斑块，斑块的数量通常为20～30个。鞍形斑块间被黄色或白色窄斜斑相隔。

A specimen from Vladivostok, Russia (Matthew Most).

Distribution: Assuming *E. anomala* is not a subspecies of *E. schrenkii*, the distribution within Greater China is Heilongjiang, Inner Mongolia, Jilin, and Liaoning. Elsewhere, it is found in extreme eastern Mongolia, the Korean peninsula, and Russia. Due to the unsettled nature surrounding the taxonomy on *E. anomala* and *E. schrenkii*, I am providing a distribution map of both species / subspecies. The green distribution represents what is thought to be *E. anomala;* the blue distribution represents what is thought to be *E. schrenkii*.

地理分布：假设赤峰锦蛇并非棕黑锦蛇的亚种，棕黑锦蛇在中国分布于黑龙江、内蒙古、吉林和辽宁。中国以外，本种分布于蒙古东部、俄罗斯和朝鲜半岛。考虑到赤峰锦蛇和棕黑锦蛇在分类上仍有诸多不确定性，本书将提供两者的分布图。绿色被认为是赤峰锦蛇的分布区域，蓝色则被认为是棕黑锦蛇的分布区域。

Habits and habitats: Diurnal and semi-arboreal. An inhabitant of the temperate montane forests within its range. This species is very docile by nature, and even wild specimens rarely bite when picked up.

习性与栖息地：白昼活动，半树栖。栖息于山林之中。生性温顺，即使野生个体被抓取也很少扑咬。

Diet and feeding habits: A powerful constrictor, primarily feeding on mammals with a preference for rodents in captivity. In the wild, this species is known to also eat both birds and bird eggs.

食性与喂养：能强有力地绞杀猎物。人工饲养个体主要以哺乳动物为食，尤其偏爱啮齿类。野生个体也会捕食鸟类和鸟卵。

Representative habitat for the species.

Captive care: See the general care section at the introduction of this genus. In captivity, this species has been recorded to have a life span of 31 years.

人工饲养：具体情况请参阅本属介绍章节中的“人工饲养概述”版块。据记载，人工饲养下本种的寿命可达31年。

A juvenile *E. schrenkii*. Much like juvenile ratsnakes of the *Pantherophis obsoletus* complex in North America, juvenile *E. schrenkii* and *E. anomala* are essentially indistinguishable (Matthew Most).

Reproduction: Oviparous. Sexual maturity is achieved at 3-4 years. In captivity, specimens are provided with a brumation period of 10-15°C for 3-4 months. Breeding is observed in the spring following brumation. The gestation period for *E. schrenkii* is 40-60 days, with eggs being laid in June-July. Clutch size ranges from 9-21 eggs with incubation of eggs at 23-25°C for 40-60 days. This species has only been recorded and observed to lay a single clutch a year.

繁殖：卵生。于3～4年达到性成熟。人工饲养下，个体应在10～15℃下经历3～4个月的冬化期。据观察，交配在冬化后的春季进行。母蛇怀卵40～60天。6～7月产卵。一窝卵9～21枚。孵化温度23～25℃，孵化期40～60天。据记载和观察，本种一年只产卵一窝。

Taxonomy and nomenclature: *Elaphe schrenkii* and *E. anomala* continually go back and forth from being considered the same species or a subspecies of *E. schrenkii*. However, recent studies have demonstrated that based on mtDNA and microsatellite data, the South Korean population aligns with the *E. schrenkii*.

分类与命名：关于棕黑锦蛇和赤峰锦蛇究竟是两个物种，亦或赤峰锦蛇为棕黑锦蛇的亚种，这方面尚无定论。然而，最近的研究表明：基于目前的"线粒体DNA"和"微卫星"数据，韩国的种群被证实为棕黑锦蛇。

A hybrid specimen of *E. anomala* x *E. schrenkii* from northern China (W. Scott Lupien).

Conservation Status: The IUCN has not assessed this species.

保护状况：本种在《濒危物种红色名录》中被定为"未予评估"等级。

A roadkilled individual from Chungcheongbuk-do, South Korea, where the genetics, morphometrics, and certainty of the species/ subspecies is still in fluctuation.

Scientific name / 学名: *Elaphe zoigeensis* (Huang, Ding, Burbrink, Yang, Huang, Ling, Chen & Zhang 2012)

English name / 英文名称: Zoige ratsnake

Chinese name / 中文名: 若尔盖锦蛇 / ruò ěr gài jǐn shé

Etymology / 种名词源: *Zoige* refers to Zoige County, and *-ensis* = "coming from." This is referring to the type locality / "*Zoige*" 指模式产地若尔盖县（Zoige County）; "*-ensis*"，拉丁语后缀，意为"来自于"。

Terra typica / 模式产地: 20 m from the mouth of Jiangzha Hot Spring (3200 m elev., N34.204978° E102.713328°) in Zoige County, Sichuan, China / 中国，四川省，阿坝藏族羌族自治州，若尔盖县，距降扎温泉口二十米处（海拔3200米）。

Length / 体长: Maturity at 550 mm (males), 750 mm (females); hatchlings emerge around 160-240 mm / 雄性体长550毫米时达到性成熟，雌性体长750毫米时达到性成熟。刚孵化的幼蛇体长160～240毫米。

Description: The cephalic scales are as follows: supralabials 7-8, the 3rd and 4th or 4th and 5th entering orbit; infralabials 9; loreal present; preoculars 3; postoculars 2; temporals 2 + 3 (anterior temporal fused); Dorsal scale rows 21:19:17 and weakly keeled toward the midline; ventrals 205; paired subcaudals 69; anal plate divided. The coloration and pattern are distinct

Specimen from the type locality (Gernot Vogel).

in having a red-brown, "M"-shaped marking on the back of the head. The species has four longitudinal, parallel rows of spots on the dorsum extending the length of the body. Each spot typically encompasses an entire scale and is red-brown in coloration. Juveniles resemble adults.

外观描述：鳞被：上唇鳞7（2-2-3、3-2-2式）或8（2-2-4、3-2-3式）；下唇鳞9；具颊鳞；眶前鳞3；眶后鳞2；颞鳞2+3（两枚前颞鳞或合为一枚）；背鳞21-19-17行，中央若干行微弱起棱；腹鳞205；尾下鳞69对；肛鳞二分。头背有一红棕色"M"形斑纹。体背具4列由红棕色小斑点所组成的纵纹。每一个小斑点通常占一枚鳞片。小亚成体与成体在外观上相似。

Distribution: The species is known from Gansu and Sichuan provinces.

地理分布：目前已知分布于四川和甘肃。

Habits and habitats: This species is very docile by nature, and even wild specimens rarely bite when picked up.

习性与栖息地：生性温顺，即使野生个体被抓取也很少扑咬。

Diet and feeding habits: Preys primarily on mammals and birds. Juveniles likely feed on insects as their small size may prove challenging for predation on mammals or birds.

Habitat shot of *E. zoigeensis* in Baozuo, Zoige, Sichuan (Wang Xiaohe 王小荷).

食性与喂养：主要以哺乳动物和鸟类为食。由于体型较小，小亚成体或以昆虫为食。

Captive care: See the general care section at the introduction of this genus. Care is very similar to that of *E. dione*.

人工饲养：请参阅本属介绍章节中的“人工饲养概述”版块，具体细节与白条锦蛇的饲养需求相似。

Reproduction: Oviparous. A limited number of facilities have kept or bred this species in captivity. Sexual maturity has been reported to be 3-4 years. In captivity, specimens are provided with a brumation period of 10-15°C for 3-4 months. Breeding is observed in the spring following brumation. The gestation period for *E. zoigeensis* is 40-60 days, with eggs being laid in June-July. Clutch size ranges from 4-12 eggs with incubation of eggs at 23-25°C for 50-70 days. Hatchlings have been reported to be difficult to establish in captivity and may require a short brumation period following hatching for individuals to start feeding.

繁殖：卵生。若尔盖锦蛇的人工饲养尚不多见。已知于3～4年达到性成熟。人工饲养下，个体应在10～15°C下经历3～4个月的冬化期。据观察，交配在冬化后的春季进行。母蛇怀卵40～60天。6～7月产卵。一窝卵4～12枚。孵化温度23～25°C，孵化期50～70天。据报道，本种幼蛇的饲养难度较大。一些刚孵化的个体在出壳后或不会立即进食，此时应短暂降温以刺激其内分泌系统。2～3周的低温期后（温度与冬化时的温度一致），个体往往会开始进食。

A reddish phase specimen (Gernot Vogel).

Taxonomy and nomenclature: There are no known subspecies.

分类与命名：目前无已知的亚种。

Comments: This species is considered Least Concern (LC) by the IUCN.

保护状况：本种在《濒危物种红色名录》中被定为"无危"等级。

Similar species: This species is distinguished from all other species of *Elaphe* by several sets of characteristics involving color and scutellation. This species could easily be confused with *E. dione* but is distinguished from other species based upon several sets of characteristics. This species has three preocular scales, which distinguishes from all other species (Schulz, 1996).

相似物种：若尔盖锦蛇可通过体色和鳞被等方面的特征同锦蛇属其它成员相区分。外观上与白条锦蛇十分相似。本种眶前鳞为三枚，这点与其它蛇种有所不同（Schulz，1996）。

Specimen from the type locality (Gernot Vogel).

The former *Orthriophis* group / 晨蛇属

As this was a fairly recent taxonomic change, I felt it would be useful to group this former complex together in the organization of the species in this book.

Etymology: The generic name is derived from the Greek *orthros* (dawn) and *ophios* (snake), referring to the snakes' high activity level in the early morning.

本属已不再有效，属下所有成员已归隶锦蛇属。考虑到相关分类变化是在近年发生，故本书将晨蛇属过去所辖的蛇种合并成一个单元进行讨论。

属名词源：源自希腊语，“*orthros*”意为“黎明”，“*ophios*”意为“蛇”。该属名意指所辖蛇种在黎明拂晓时有很高的活跃度（该属有时也被称作“曙蛇属”）。

General captive care / 人工饲养概述：

If available, captive-bred specimens should be acquired. Wild-caught or farm-raised specimens typically have parasites, are malnourished, and are heavily stressed, resulting in a weakened immune system. If working with wild-caught or farm-raised individuals, a qualified veterinarian should conduct a fecal and lung wash to test for parasites. Animals should be quarantined according to the protocols set out by the facility or the keeper, but an approximate quarantine period of 3 months to 1 year is advised to prevent the spread of any potential diseases or parasites.

野外捕获个体及部分养殖场所提供的个体或因寄生虫感染、营养不良、应激反应等因素存在免疫力低下的隐患。若购入以上个体，建议联系有资质的兽医对个体的呼吸道和消化道样本进行寄生虫检测。若可能，应从可靠渠道获取人工繁殖个体。新入手的个体需实行隔离，隔离时长可根据提供方给出的建议。一般建议隔离3个月至1年。

This section is the generalized care for all members in the former *Orthriophis* genus; for individual species specifics, please look at the Captive Care section under the species of interest. These are all large and very active species that do well in both an arboreal and terrestrial setup. Individuals will use vertical climbing space if provided with multiple basking areas. These species are kept at temperatures that range from 21-25°C and relative humidity of 60-70%. Adults and juveniles are housed in enclosures with cypress mulch bedding. Adults and juveniles should be offered a minimum terrarium of 86 x 46 x 16 cm. Hatchlings can be housed in a plastic shoebox or commercialized snake breeding containers measuring 31 x 19 x 10 cm for their first two years of life before they should be transitioned to a larger terrarium. Paper towels are used as a substrate for hatchlings, as this genus has a very strong feeding response and is used to avoid the accidental ingestion of substrate material. Hatchling setups should also include a moist hide which contains sphagnum moss, which is commonly used when the juveniles are opaque. Some neonates may not start feeding immediately after hatching and should be offered a short cooling period to stimulate the endocrine system. After a brief period of 2-3 weeks at cooler temperatures, similar to those of brumation, these individuals will usually start feeding.

Similar to the other species, these animals are also cycled with weather patterns. These animals are maintained at temperatures of 10-15°C for 3-4 months. To induce breeding behavior, males should be introduced to females following brumation.

本版块的饲养方法适用于之前归隶晨蛇属的蛇种。欲知更多细节，请参阅各蛇种介绍章节中的“人工饲养”版块。本属蛇种体型较大，生性活跃。环境布置方面，无论是以树栖还是地栖为主，本属成员皆可很好适应。若饲养场地设置有多个晒点，它们会充分利用环境中的垂直空间。饲养环境的温度应为21～25℃，相对湿度介于60～70%。成体及小亚成体的饲养容器应不小于86 x 46 x 16厘米，底材可选用柏树皮。幼体可饲养在普通塑料盒或专业饲养箱内，空间大小建议在31 x 19 x 10厘米。不过随着幼体的成长，两年后应考虑为其提供更大的饲养空间。考虑到本属蛇种有较强的摄食反应，为避免误食底材，幼体的底材建议选用厨房纸。幼体的环境中应放置加湿躲避穴，可在其中填充泥炭藓用于提高湿度。小亚成体蜕皮时也常使用该类躲避穴。部分刚孵化个体在出壳后或不会立即进食，此时应短暂降温以刺激其内分泌系统。2～3周的低温期后（温度与冬化时的温度一致），个体往往会开始进食。为模拟本属蛇种在自然环境中的节律，个体应经历为期3～4个月的冬化，冬化时的温度应保持在10～15℃。为诱发繁殖行为，公母个体应在冬化后置于同一饲养环境中。

Egg Incubation: Eggs of this genus are significantly different from other species and should be incubated in a larger container to allow for proper air circulation. Several breeders have documented animals that developed to full term only to find that the embryos had drowned. This is hypothesized to be a result of improper egg incubation. Factors that could be contributing to this death could be a combination of the eggs not having proper ventilation or air circulation resulting in excessive moisture in the incubation chamber in combination with temperatures that are too hot. Eggs of this genus should be housed in commercially available incubators set to 25-27°C or room temperature and measuring a minimum of 31 x 19 x 10 cm. Still, this size may need to be increased depending on the number of eggs. Incubating at room temperature will result in a hatching delay by approximately 10-15 days.

One way of suspending eggs above vermiculite substrate (Ryan Goodman and Bejamin Levin, R & B Reptiles).

卵的孵化：同其它蛇种的卵相比，本属成员的卵有较大不同。孵化盒尺寸应更大以促进空气的流通。不少本属蛇种的饲养者都遇到了胚胎发育至后期溺毙的情况。据推测，这是由于孵化不当所造成的，包括孵化温度过高以及孵化盒内空气流通不畅所导致的湿度过大。使用专业孵化器以25～27℃或室温进行孵化皆可使本属蛇种的幼体顺利出壳，但室温下的孵化期要多10～15天。孵化器的尺寸建议不小于31 x 19 x 10厘米（若卵的数量较多，则需更大尺寸）。

The containers must be well ventilated with holes on all sides to allow for fresh air flow and remove excessive moisture. Excessive moisture and improper ventilation can create issues for developing embryos/eggs, as mold and fungus can grow on the egg's surface. Eggs can be incubated upon a variety of media, but preference is vermiculite or perlite. Both substrates retain moisture and aid in maintaining humidity. Still, both media can dry out and may require additional water if necessary. To achieve proper moisture for incubating, dry media should be measured to have a 5 cm depth within the container. A 2:1 ratio of substrate to water by weight is normally sufficient to ensure proper moisture and humidity during incubation. One can check the appropriate balance by squeezing the mixture by hand. Expressing water means the water content is too high. After achieving the proper moisture, the media, followed by the eggs, can be placed in the incubator. When eggs are laid, it is best to gently separate eggs to allow for adequate airflow within the incubation chamber. Eggs are placed atop the media, and a means of suspending the eggs over the media is used for this genus to avoid undue moisture uptake. As a precaution, eggs should not be buried in the substrate for this genus.

Clutch of *E. moellendorffi* hatching (Matthew Most).

盛放卵的孵化盒四周均应有孔，以保证空气的流通及多余湿气的排出。湿度过高或通风不畅将对卵的发育造成影响，易令卵壳表面滋生真菌或霉菌。孵化介质有多种选择，推荐使用蛭石或珍珠岩。两者皆有不俗的锁水性，有利于湿度的保持。然而，两种介质也可能在孵化的过程中变得过干。若有必要，需另行补水。为确保孵化过程中湿度始终处于理想水平，孵化介质（干燥状态下）在容器内的铺设厚度需达到5厘米。然后，水和孵化介质在重量上以1:2进行配比。一种检验配比是否合适的有效方法是用手捏兑水后的孵化介质。若有水渗出，则表明水份过高。放置卵时，建议互相之间保持一定的距离，以促进孵化盒内空气的流通。卵可放置在孵化介质表面或悬于孵化介质之上。第二种方法可有效降低因湿度过高对孵化造成的不利影响。切不可将本属蛇种的卵埋入孵化介质中进行孵化。一切准备妥帖之后，孵化盒放入孵化器中。

Euprepiophis mandarinus specimens hatching from a suspended media (Ralph Polinski Midwest Serpentarium).

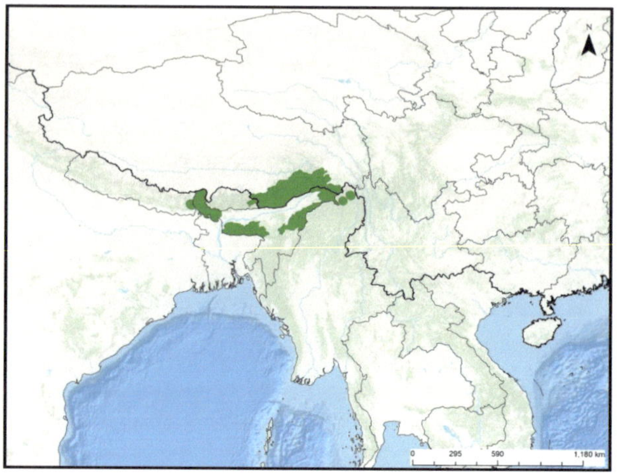

Scientific name / 学名: *Elaphe* [formerly *Orthriophis*] *cantoris* (Boulenger, 1894)

English name / 英文名: Himalayan ratsnake, Eastern trinket snake

Chinese name / 中文名: 坎氏锦蛇 / kǎn shì jǐn shé

Etymology / 种名词源: The specific name, *cantoris* is named in honor of Danish zoologist Theodore Edward Cantor (1809-1860), a surgeon of the East Indian Company at Calcutta, a naturalist, and herpetologist for British India / 以丹麦动物学家、博物学家Theodore Edward Cantor（1809-1860）的名字来命名。他曾是东印度公司驻加尔各答的一名外科医生，也是一名两栖和爬行动物学家。

Terra typica / 模式产地: Cherrapunji, East Khasi Hills, Meghalaya State, India（N25.3° E91.7° ）/ 印度，梅加拉亚邦，东卡西丘陵，乞拉朋齐

Length / 体长: Maturity at 1400-1600 mm; Hatchlings emerge around 300-380 mm. Maximum size 1960 mm / 体长1400～1600毫米时达到性成熟。刚孵化的幼蛇体长300～380毫米。已知最大个体全长1960毫米。

Description: The cephalic scales are as follows: Supralabials 8 or 9 (4th, 5th or 3-5 touching eye); last supralabial below eye not touching temporals; one large preocular; often with a subocular and sometimes an additional presubocular; postoculars 2; temporals 2+2 or 2+3; ventrals 212-236, subcaudals 65-88, paired; anal plate usually undivided. Ventrals angled at the sides. Loreal scale a little longer than high. Scales smooth or weakly keeled with dorsal scales in about 19-21:19-21:17 rows. Head elongate, slightly broader than neck; iris often red. Forebody grayish or yellowish-brown, marked with a series of single or paired, dark or reddish-brown oval or squarish blotches on midback; side of the body has smaller spots that alternate with those on the back; some scale edges white—the underside of head yellow to orange. Ventral scales initially yellowish, bluish-gray towards the tail; the underside of the tail sometimes pink. Scales on the belly and under tail sometimes spotted with brown or black. Occasionally underside entirely dark brown or black. Colors and patterns dull on older snakes. Juveniles resemble adults.

Specimen from India (Ashok Captain).

外观描述: 鳞被:上唇鳞8（3-2-3、2-3-3式）或9（3-2-4、2-3-4式）；眶下最后1枚上唇鳞不切颞鳞；眶前鳞较大，1枚；常有眶下鳞1，个别另有眶前下鳞1；眶后鳞2；颞鳞2+2或2+3；腹鳞212～236，具侧棱；尾下鳞65～88对；肛鳞通常完整；颊鳞长略大于高；背鳞平滑或微弱起棱，颈部19～21行，中段19～21行，肛前17行。头部修长，略宽于颈。眼睛虹膜通常为红色。体前段背面呈棕黄或浅灰色。正背具一条由一系列斑块所组成的纵纹。斑块成单或成双排列，颜色为黑色或红褐色，形状为椭圆或方形。体两侧也各具一条由斑块所连成的纵纹，其上的斑块与正背的斑块交错排列。部分鳞片镶有白边。头腹面黄或橙色。体前段腹面淡黄色，体后段腹面趋于蓝灰色；个别尾腹面粉色。偶有个体整个腹面呈深棕色或黑色。个别躯尾腹面散有棕色或黑色斑块。小亚成体与成体在外观上相似，不过年轻个体颜色及斑纹更为艳丽。

Distribution: Within Greater China, this species is found in Xizang = Tibet. Outside of Greater China, it has been recorded in India, Nepal, Myanmar, and Bhutan. Found in elevations between 1000 to 2300 m.

地理分布: 在中国，分布于西藏。中国以外，分布于印度、尼泊尔、缅甸和不丹。垂直分布1000～2300米。

Habits and habitats: Poorly known in the wild. Mainly active in the early morning. This species has been found in dense primary montane forests, secondary forests, near houses, and agricultural fields. Mostly terrestrial.

习性与栖息地: 对本种在野生环境中的情况知之甚少。主要在清晨活跃，地栖性为主。常见于茂密的原始山林、次生林、人类住所及农田附近。

Habitat of *E. cantoris* (Ashok Captain).

Diet and feeding habits: Captive specimens feed on rodents, killing via constriction, and likely feed on other mammals and birds within their natural range.

食性与喂养：人工饲养个体主要以啮齿类为食，有绞杀猎物的习性。自然环境中，或还捕食其它哺乳动物和鸟类。

Captive care: See the general care at the introduction of this group (the *Orthriophis* group). In captivity, the longevity of this species has been recorded to at least 22 years (R. Fesser, pers. comm.).

人工饲养：具体情况请参阅本属介绍章节中的“人工饲养概述”版块，据记载，人工饲养下本种的最长寿命至少可达22年（R. Fesser，个人通讯）。

Freshly hatched juveniles (left); pair of adults copulating (right) (both images Rainer Fesser).

Reproduction: Oviparous. Males attain larger sizes than females. Sexual maturity is achieved at 3-4 years. In captivity, specimens are provided with a brumation period of 10-15°C for 3-4 months. Breeding is observed in the spring shortly after the following brumation. The gestation period for *cantoris* is 60 days, with eggs being laid in June and July. Clutch size ranges from 5-10 eggs with incubation of eggs at 23-25°C for 80-110 days. This species has only been recorded and observed to lay a single clutch a year.

繁殖：卵生。雄性体型大于雌性。3～4年达到性成熟。人工饲养下，个体应在10～15°C下经历3～4个月的冬化期。据观察，交配在冬化后的春季进行。母蛇怀卵60天左右。6～7月产卵。一窝卵5～10枚。孵化温度23～25°C，孵化期80～110天。据记载和观察，本种一年只产卵一窝。

Taxonomy and nomenclature: Originally *Elaphe*, then reassigned to *Orthriophis* in 2010, and then assigned back to *Elaphe* in 2018.

分类与命名：最初为锦蛇属成员，2010年改隶晨蛇属，2018年回归锦蛇属。

Similar species: *E.* [*O.*] *cantoris* has many similarities to *E.* [*O.*] *taeniura,* including the anatomy of their eggs. The eggs of this species do not have a thick wall like that of *E.* [*O.*] *moellendorffi.* This species has a thin-walled eggshell, and specimens cut the eggs as commonly observed in *E.* [*O.*] *taeniura.*

相似物种：坎氏锦蛇同黑眉锦蛇有许多相似之处，包括卵的构造。同百花锦蛇的卵卵壳较厚不同，坎氏锦蛇的卵卵壳较薄。幼蛇出壳时从卵的侧面破壳而出，这点也和黑眉锦蛇相一致。

Conservation Status: The IUCN has not assessed this species.

保护状况：本种在《濒危物种红色名录》中被定为"未予评估"等级。

Example of *E. cantoris* habitat (Scott Trageser).

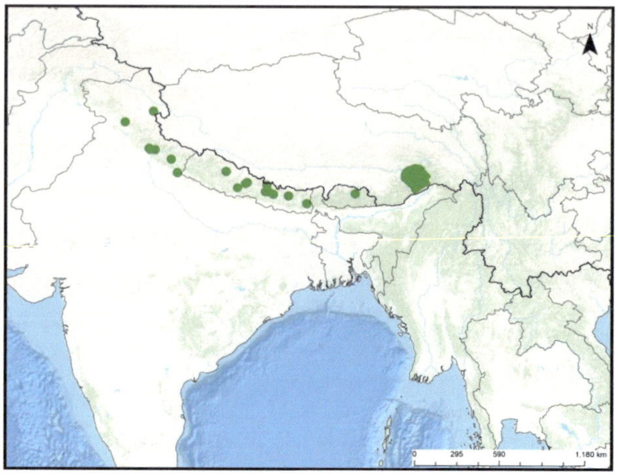

Scientific name / 学名: *Elaphe [Orthriophis] hodgsonii* (Günther, 1860)

English name / 英文名: Hodgson's ratsnake, Himalayan Trinket Snake

Chinese name / 中文名: 南峰锦蛇 / nán fēng jǐn shé

Etymology / 种名词源: The specific name *hodgsonii* is in honor of British naturalist Brian Houghton Hodgson (1801—1894) / 以英国博物学家Brian Houghton Hodgson (1801-1894) 的名字来命名。

Terra typica / 模式产地: Lake Moriri, Ladakh, India (N32.883° E78.35°) / 印度，拉达克，莫里里湖。

Length / 体长: Maturity at 1200-1600 mm; Hatchlings emerge around 300-400 mm. Maximum males 1420 mm (270 mm tail), female 1500 mm (310 mm tail) / 体长1200～1600毫米时达到性成熟。刚孵化的幼蛇体长300～400毫米。已知最大雄性个体全长1420毫米，其中尾长270毫米；最大雌性个体全长1500毫米，其中尾长310毫米。

Description: The cephalic scales are as follows: supralabials 8 rarely 9 (3rd and 4th, 4th and 5th, 3-5, or 4-6 in contact with orbit); last supralabial below eye not touching temporals; one large preocular; rarely a small subocular in front of preocular; postoculars 2; temporals 2+2 or 2+3, rarely 3+3, 1+3 or 2+4. Dorsal scale rows in 21 or 23:23:17 rows. Ventrals 228-247 with a strong lateral keep; anal scale divided; subcaudals 72-92, paired. Slender-bodied; scales smooth or weakly keeled. Elongated head slightly broader than neck; pupil round. Yellowish to olive-brown back; many scales edged with black or white, grouped form cross-bands or a mesh-like pattern. Sometimes a central dark blotch on top of the head. Labials and underside yellowish. Outer edges of belly scales distinctly keeled sometimes edged or spotted with black. The underside of the tail is reddish-brown to black. Some specimens from Nepal are gray, with an indistinct brown stripe on each side of the body. Juveniles may resemble adults; they may be gray with thin black lines or have the midbody marked with four rows of reddish-brown blotches that change into four black lines towards the tail.

Adult female *E. hodgsonii* with a recent clutch of eggs (Rainer Fesser).

外观描述：鳞被：上唇鳞8（2-2-4、3-2-3、2-3-3或3-3-2式），少数9（2-2-5、3-2-4、2-3-4或3-3-3式）；眶下最后1枚上唇鳞不切颞鳞；眶前鳞较大，1枚；少数另有1枚较小的眶前下鳞；眶后鳞2；颞鳞2+2或2+3，少数3+3、1+3或2+4；背鳞21（23）-23-17行，鳞片平滑或微弱起棱；腹鳞228～247，具侧棱；肛鳞二分；尾下鳞72～92对。体型修长。头部较长，略宽于颈。瞳孔圆形。体背淡黄或黄褐色，不少鳞片边缘黑或白色，形成横斑或网状斑纹。部分个体头背具一黑色斑块。头躯腹面淡黄色。部分个体腹鳞边缘黑色。尾腹面红褐或黑色。部分采自尼泊尔的标本背面灰色，体两侧各具一条不明显的褐色纵纹。部分小亚成体与成体在外观上相似，部分则略有差异：有些小亚成体背面灰色，体背具黑色细纹；有些小亚成体体中段具4条由红褐色斑块所组成的纵纹，斑块在体后段逐渐汇为4条黑线。

Distribution: Within Greater China, this species is found in Xizang = Tibet. Elsewhere outside of China, it is found in Bhutan, India, and Nepal. Found from 835-3200 m elev.

地理分布：在中国，分布于西藏。中国以外，分布于不丹、印度和尼泊尔。垂直分布835～3200米。

Habits and habitats: Poorly understood. Diurnal. A high elevation species found in moist deciduous forests, moist oak forests, dry coniferous forests, near urban habitation, and edges of agricultural fields close to water.

General habitat of *E. hodgsonii* in Nepal (Dan Rosenberg).

习性与栖息地:对此知之甚少。白昼活动,栖息于高海拔地区。常见于潮湿的落叶林、潮湿的橡树林、干燥的针叶林、人类住所附近及水源周边的农田。

Diet and feeding habits: This species is known to feed on small mammals, including rodents. *E. hodgsonii* may also be an opportunistic feeder during seasonal change, feeding on birds, bird eggs, skinks, and amphibians.

食性与喂养:已知以小型哺乳动物为食,包括啮齿类。也可能为机会主义者,在不同的季节猎食不同的猎物,包括鸟类、鸟卵、石龙子以及两栖类。

Captive care: See the general care at the introduction of this group (the *Orthriophis* group).

人工饲养:具体情况请参阅本属介绍章节中的"人工饲养概述"版块。

Reproduction: Oviparous. Sexual maturity is achieved at 3-4 years. In captivity, specimens are provided with a brumation period of 10-15℃ for 3-4 months. Breeding is observed in the spring shortly after the following brumation. The gestation period for *E. hodgsonii* is 60-70 days, with eggs being laid in June and July. Clutch size ranges from 5-10 eggs with incubation of eggs at 23-25℃ for 70-80 days. A captive specimen laid and guarded six eggs that hatched in about 130 days. This species has only been recorded and observed to lay a single clutch a year.

繁殖:卵生。3~4年达到性成熟。人工饲养下,个体应在10~15℃下经历3~4个月的冬化期。据观察,交配在冬化后的春季进行。母蛇怀卵60~70天。6~7月产卵。一窝卵5~10枚。孵化温度23~25℃,孵化期70~80天。曾有一人工饲养的个体产卵6枚并护卵约130天直至幼蛇最后出壳。据记载和观察,本种一年只产卵一窝。

A newly hatched *E. hodgsonii* (Rainer Fesser).

Taxonomy and nomenclature: There are no known subspecies. This species was originally in the genus *Elaphe* but was reassigned to *Orthriophis* in 2010, but in 2017 it was assigned back to *Elaphe*.

分类与命名：目前无已知亚种。本种最初为锦蛇属成员，2010年改隶晨蛇属，2017年回归锦蛇属。

Conservation Status: The IUCN has not assessed this species.

保护状况：本种在《濒危物种红色名录》中被定为“未予评估”等级。

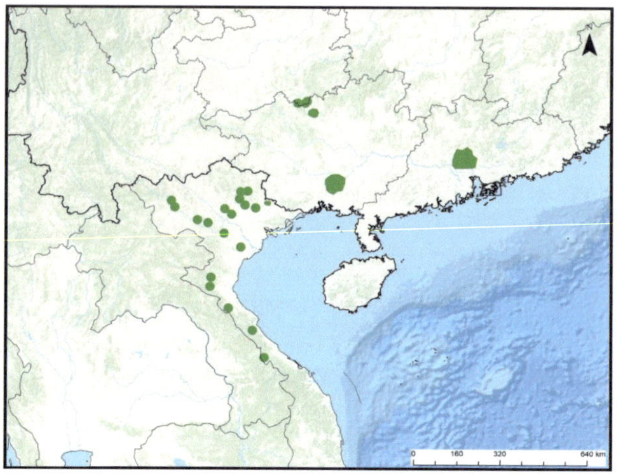

Scientific name / 学名: *Elaphe* [*Orthriophis*] *moellendorffi* (Boettger, 1886)

English name / 英文名: Moellendorff's ratsnake, 100 Flower ratsnake, Red-Headed Ratsnake

Chinese name / 中文名: 百花锦蛇 / bǎi huā jǐn shé

Etymology / 种名词源: The specific name, *moellendorffi* is named after Otto Franz von Möllendorff, a German malacologist (1848 - 1903) / 以德国软体动物学家Otto Franz von Möllendorff (1848 - 1903) 的名字来命名。

Terra typica / 模式产地: Yong Jiang (= River) near Nanning, Guangxi, China (N22.82° E108.36°) / 中国，广西壮族自治区，南宁市，邕江。

Length / 体长: Large bodied snake exceeding 2 meters. Maturity at 1200-1400 mm; Hatchlings emerge at 370-490 mm. Mell (1922) records a Guangdong specimen measuring 2150 mm / 大个体的百花锦蛇体长可超过2米。体长1200～1400毫米时达到性成熟。刚孵化的幼蛇体长370～490毫米。Mell (1922) 曾记载一体长达到2150毫米的个体，该蛇采集自广东。

Description: The cephalic scales are as follows: supralabials 9-10, somewhat quadrangular, 5th and 6th, or 6th and 7th in contact with orbit; infralabials 11-12; loreal 1, twice as long as high; 1 very large preocular, larger than the eye; distinct presubocular, placed above the 5th supralabial. The dorsal scales are moderately keeled and in 25-27 scale rows at midbody. There are 268-274 strongly angulated ventrals and 97-99 subcaudals. The anal plate is divided. On a grayish background are about 28-32 large, irregular reddish-brown saddles with black edges. Most of the saddles are vaguely rectangular, but some may be broken and have multiple corners. On the lower side is a series of smaller reddish-brown blotches. Scattered scales along the sides are edged with dark. The belly is yellowish-white with black checks. The head and neck are unmarked, the head itself large, with a long snout and very distinct neck. The head's entire upper surface is bright red to coppery red, shading to gray on the neck and white on the labial scales. The tail is circled with brownish-gray bands on a pinkish background to bright red, presenting a red-ringed effect. Juveniles resemble adults.

外观描述：鳞被：上唇鳞9（4-2-3、5-2-2式）或10（4-2-4、5-2-3式），形状略成四边形；下唇鳞11～12；颊鳞1，长两倍于高；眶前鳞很大，1枚，尺寸大于眼；另有一枚明显的眶前下鳞，位于第五枚上唇鳞之上；体中段背鳞25～27行，适度起棱；腹鳞268～274，侧棱明显；尾下鳞97～99对；肛鳞二分。背面浅灰色，中央具28～32个呈不规则鞍形的红褐色斑块，斑块边缘黑色。体两侧近腹面另各有一行红褐色斑块，体侧斑块小于正背斑块。体侧部分鳞片镶有黑边。腹面黄白色，具黑色方斑。头、颈部无斑纹。头较大，与颈区分明显，吻长。头背为鲜红或铜红色，颈部渐为灰色，唇部白色。尾背面淡粉或鲜红色，具若干棕灰色环斑。小亚成体与成体在外观上相似。

Distribution: Within Greater China, this species is found in Guangdong, Guangxi, and Guizhou. Elsewhere, it is found in northern Vietnam. Found from 15 - 500 m.

地理分布：在中国，分布于广东、广西和贵州。中国以外，分布于越南北部。垂直分布15～500米。

General habitats of *E. moellendorffi* in central Vietnam (top image: author; bottom image: Mike Pingleton.

Habits and habitats: This species is cathemeral and frequently found in limestone caves. Also found in rocky, forested hillsides, meadows, and bamboo thickets.

习性与栖息地：昼夜皆会活动。常发现于石灰岩洞穴，也见于多岩石的森林坡地、草场以及竹林。

Diet and feeding habits: Primarily feeds on mammals with a preference for rodents in captivity. In the wild, this species is known to live in limestone caves occupied by bats, which this species is known to feed on.

Specimen searching the ceiling for bats in a karst cave in Dong Thien Ha in Trang An, Ninh Binh, Vietnam (Thorsten Kahlert).

食性与喂养：人工饲养个体主要以哺乳动物为食，尤其偏爱啮齿类。自然环境中，常栖息于有蝙蝠生存的石灰岩洞穴并捕食蝙蝠。

Captive care: See the general care at the introduction of this group (the *Orthriophis* group). Hatchlings may require brumation before they begin feeding as they typically hatch in November/December. In captivity, this species has been recorded to have a life span of 15-20 years.

人工饲养：具体情况请参阅本属介绍章节中的"人工饲养概述"版块。幼蛇通常于11～12月孵化出壳，故进食往往在冬化之后。据记载，人工饲养下本种的寿命为15～20年。

Reproduction: Oviparous. Sexual maturity is achieved at 5-6 years. In captivity, specimens are provided with a brumation period of 10-15°C for 3-4 months. Breeding is observed in the spring following brumation. The gestation period for *E. moellendorffi* is 50-60 days, with eggs being laid in July-August. Clutch size ranges from 5-10 eggs with incubation of eggs at 23-25°C for 80-110 days. This species has only been recorded and observed to lay a single clutch a year. By contrast to other species, eggs of this species are highly calcified and hard, unlike *cantoris, hodgsonii,* or *taeniura* eggs, which have a leathery texture and consistency. Hatchlings emerge from the eggs' polar ends, puncturing the ends of the eggs rather than slitting the lateral shell with their egg tooth during hatching.

Above: Juvenile *E. moellendorffi* emerging from the polar end of an egg (Matthew Most).
Below: An empty *E. moellendorffi* egg, showing the polar end broken through (Nathan Shepard).

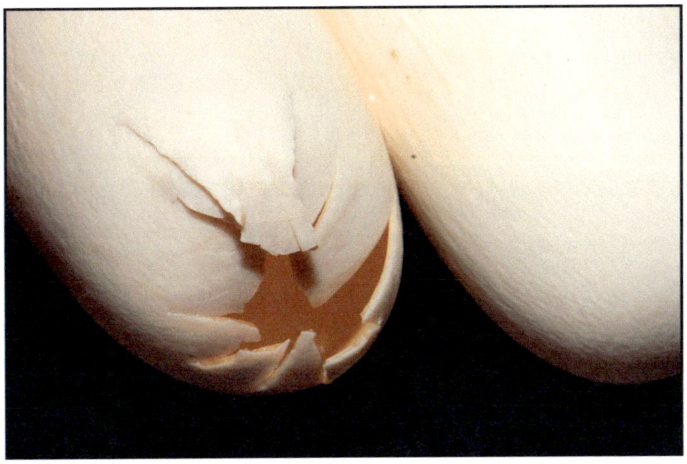

Left: Clutch of *E. moellendorffi* hatching. Right: A hypomelanistic *E. moellendorffi* (Matthew Most, both images).

繁殖：卵生。5~6年达到性成熟。人工饲养下，个体应在10~15℃下经历3~4个月的冬化期。据观察，交配在冬化后的春季进行。母蛇怀卵50~60天。7~8月产卵。一窝卵5~10枚。孵化温度23~25℃，孵化期80~110天。据记载和观察，本种一年只产卵一窝。百花锦蛇的卵卵壳高度钙化且坚硬，与坎氏锦蛇、南峰锦蛇和黑眉锦蛇的卵卵壳呈皮革质感完全不同。出壳时，百花锦蛇的幼蛇从卵的两端，而非侧面破壳而出。

Taxonomy and nomenclature: There are no subspecies.

分类与命名：目前无已知亚种。

Conservation Status: This species is listed as Vulnerable on the IUCN Redlist, last assessed in 2011.

保护状况：本种在《濒危物种红色名录》中被定为“易危”等级（最近一次评定为2011年）。

Aberrant color morphs with reduced patterns (Matthew Most).

Comments: The species is known to naturally hybridize with *E. taeniura.* This species of ratsnake has a very rigid and hard eggshell. Hatchlings emerge from the polar ends when breaking through their shell, which is not typical of other Asian species aside from *Ptyas.* Supralabials 5 and 6 entering the eye is also a feature of the genus *Gonyosoma.* Unfortunately, this species is frequently seen in meat markets in southern China.

注释：已知本种在野生环境中会同黑眉锦蛇自然杂交。百花锦蛇的卵卵壳坚硬，幼蛇出壳时从卵的两端破壳而出，这一特点在除鼠蛇属（*Ptyas*）以外的亚洲蛇种中并不多见。百花锦蛇第5和第6枚上唇鳞入眶，这一特征则是树栖锦蛇属蛇种的特点。不幸的是，百花锦蛇经常出现在华南地区的餐桌之上。

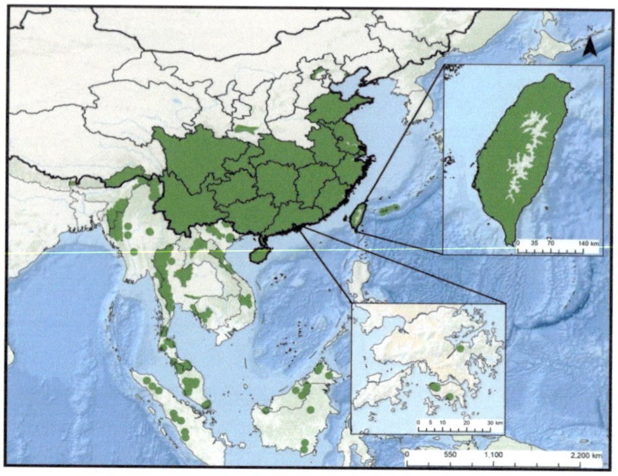

Scientific name / 学名: *Elaphe* [*Orthriophis*] *taeniura* (Cope, 1861)

English name / 英文名: Beauty snake, beauty ratsnake, cave racer, stripe-tailed racer

Chinese name / 中文名:黑眉锦蛇 / hēi méi jǐn shé

Etymology / 种名词源: Gr. *taenia* = ribbon or band, and Gr. *–ura* = tail. This is referring to the pattern on the posterior portion of the body / 源自希腊语中的"*taenia*"一词，意为"带、环"；"*-ura*"，希腊语后缀，意为"尾巴"。该种名意指黑眉锦蛇体后段的斑纹。

Terra typica / 模式产地: Ningbo, Zhejiang, China (N29.87° E121.54°) / 中国，浙江省，宁波市

Length / 体长: Maturity at 1200-1600 mm; Hatchlings 340-480 mm. Maximum 2700 mm (the *E. t. friesi* subspecies reaches this size) / 体长1200～1600毫米时达到性成熟；刚孵化的幼蛇体长340～480毫米。已知最大个体体长2700毫米（*E. t. friesi*亚种可生长至该尺寸）。

Description: The cephalic scales are as follows: Supralabials 8-9, rarely 7 or 10 (4,5 or 5,6 in contact with orbit); last labial below eye not touching temporals; infralabials 10-13; preocular 1; subocular sometimes present; postoculars 2, rarely 3; the combinations of anterior and posterior temporals, respectively, greatly vary: 2+2, 2+3, 3+3, rarely 3+2, 2+4, 4+4, or 3+4. The dorsal scales in 23: 23 or 25: 17 or 19 rows, which are feebly keeled except on the sides, which are smooth. Ventrals 236-260, with a strong lateral keel; anal scale divided; paired subcaudals 89-120. The head is elongate; the body is moderately heavy or robust; the tail is long. The eye is medium to large; iris is olive-gray to a dirty yellow, and a broad gray-black to black horizontal band runs across the eye and appears as a prominent postocular stripe. The upper head is olive-yellow to dark olive-gray and may be tinged with brown; the labials are lighter. The upper body's general color is olive-yellow to yellow gray tinged with varying shades of green and black. The ventral side of the head and neck are white or cream-colored. The rest of the ventral

E. t. yunnanensis (left) from western Hubei, and *E. t. taeniura* (right) from Fujian.

body and tail are yellow, with a few small irregular gray or gray-black designs anteriorly followed by more numerous and progressively darker patterns posteriorly. The dorsal pattern varies tremendously, depending on the subspecies. Juveniles resemble adults.

外观描述：鳞被：上唇鳞8（3-2-3、4-2-2式）或9（3-2-4、4-2-3式），少数7（3-2-2、4-2-1式）或10（3-2-5、4-2-4式）；眶下最后1枚上唇鳞不切颞鳞；下唇鳞10～13；眶前鳞1；个别具眶下鳞；眶后鳞2，少数3；颞鳞变化较大，通常2+2、2+3或3+3，少数3+2、2+4、4+4或3+4；背鳞23-23（25）-17（19）行，除最外数行平滑外，均微弱起棱；腹鳞236～260，侧棱明显；肛鳞二分；尾下鳞89～120对。头部较长，躯体较粗壮，尾长。眼中等偏大，眼虹膜为橄榄灰或橘黄色。一道粗黑横纹贯穿眼部。头背橄榄黄或灰褐色。唇部色淡。体背橄榄黄、黄灰或略呈深绿色。头、颈部腹面白色或米色。躯尾腹面黄色，体前段具不规则灰色或灰黑色小斑块，体后段斑块数量增多且颜色渐深。不同亚种间体背的斑纹有较大差异。小亚成体与成体在外观上相似。

Distribution: Within Greater China, this species is one of the more ubiquitous species, found nearly countrywide, except for the deserts in the west. It is found in Anhui, Chongqing, Fujian, Hainan, Hebei, Henan, Hubei, Hunan, Gansu, Guangdong, Guangxi, Guizhou, Jiangsu, Jiangxi, Liaoning, Sichuan, Shanxi, Shaanxi, Taiwan, Tibet, Yunnan, and Zhejiang. The population in Hong Kong is in debate. See the **Comments** section for more information on the Hong Kong population. Outside of Greater China, it is found in Borneo, Cambodia, India, Japan, Laos, Malaysia, Myanmar, Nepal, Singapore, Sumatra, Thailand, and Vietnam. There is an invasive subspecies population in the Ryukyu islands of Japan (in addition to the native subspecies population). The species is found in elevations between 700 and 3100 m.

General habitat of *E. taeniura*. This species can be found in nearly all manner of habits in eastern China.

地理分布：广布种。在中国，除西部沙漠外的其它地区几乎都有分布，具体包括安徽、重庆、福建、海南、河北、河南、湖北、湖南、甘肃、广东、广西、贵州、江苏、江西、辽宁、四川、山西、陕西、台湾、西藏、云南和浙江。本种于香港的分布存在争议，详见本章节中的“注释”版块。中国以外，分布于婆罗洲（分属印度尼西亚、马来西亚和文莱三国）、柬埔寨、印度、日本、老挝、马来西亚、缅甸、尼泊尔、新加坡、苏门答腊岛（印度尼西亚西部一大型岛屿）、泰国和越南。本种于琉球群岛也有分布，详见本章节中的“分类与命名”版块。垂直分布700～3100米。

Habit and habitats: This species is primarily diurnal but can be found active at night during the hotter summer months. It can be found in grasslands, forests, lowlands, mountains, and around human dwellings, usually in proximity to water. It is quite agile, moves fast on the ground as well as in shrubs or trees. This snake is docile and adapts well to captivity. If threatened, inflates throat vertically, lifts the forebody off the ground into S-shaped coils, opens mouth, hisses, and vibrates tail, and will repeatedly strike if continued to be provoked.

习性与栖息地：白昼活动为主，盛夏夜晚也较为活跃。可见于草原、森林、低地、山区以及人类住所附近，通常于水源周边活动。行动敏捷，可在地面、灌木、甚至树上快速移动。生性温顺，对人工饲养有很强的适应性。若感知威胁，黑眉锦蛇常侧扁颈部，体前段抬起并呈“S”形卷曲，抖动尾部并张嘴发出嘶嘶声。若被持续滋扰，会反复扑咬。

Diet and feeding habits: Primarily feeds on mammals with a preference for rodents in captivity. This species is known to eat amphibians, mammals, birds, and bird eggs in the wild.

食性与喂养：人工饲养个体主要以哺乳动物为食，尤其偏爱啮齿类。自然环境中，已知捕食两栖类、哺乳类、鸟类及鸟卵。

Captive care: See the general care at the introduction of this group (the *Orthriophis* group). In captivity, this species has been recorded to have a life span of 29 years (R. Fesser, pers. comm.).

人工饲养：具体情况请参阅本属介绍章节中的“人工饲养概述”版块。据记载，人工饲养下本种的寿命可达29年（R. Fesser，个人通讯）。

Reproduction: Oviparous. Sexual maturity is achieved at 3-4 years. In captivity, specimens are provided with a brumation period of 10-15℃ for 3-4 months. Breeding is observed in the spring following brumation. The gestation period for *E. taeniura* is 40-60 days, with eggs being laid in June-July. Clutch size ranges from 2-10 eggs with incubation of eggs at 23-25℃ for 50-70 days. This species has only been recorded and observed to lay a single clutch a year.

繁殖：卵生。3～4年达到性成熟。人工饲养下，个体应在10～15℃下经历3～4个月的冬化期。据观察，交配在冬化后的春季进行。母蛇怀卵40～60天。6～7月产卵。一窝卵2～10枚。孵化温度23～25℃，孵化期50～70天。据记载和观察，本种一年只产卵一窝。

Taxonomy and nomenclature: Five subspecies (A-E) are identified among the Indochinese subspecies complex.

A) *E. t. taeniura*; Eastern China; supralabials 8-9, 5th and 6th, or 4th and 5th, or $4\text{-}6^{th}$ in contact with orbit; suboculars 1; midbody scales 23-25; ventrals 225-255; subcaudals 84-112.

B) *E. t. friesi;* Taiwan. This subspecies has been introduced (invasive) on the island of Okinawa, Japan. Suboculars 1 or 2; midbody scales 25; ventrals 243-262; subcaudals 96-123.

C) *E. t. mocquardi;* China and Vietnam. Supralabials 9, with 5th and 6th in contact with orbit. Suboculars 1; midbody scales 23-25, ventrals 251-264; subcaudals 90-125.

D) *E. t. schmackeri;* Japan (Ryukyu Islands). Suboculars 1; midbody scales 25; ventrals 246-260; subcaudals 104-125;

E) *E. t. yunnanensis*; supralabials 7-9, 4th and 5th, or 5th and 6th in contact with orbit; suboculars usually 0 or 1, rarely 2; midbody scales 23-25; ventrals 236-260; subcaudals 89-120. Bhutan, China, India, Myanmar, Thailand, and Vietnam. The population in Beijing has recently been identified as belonging to this subspecies.

F) and G): Members of the Indomalayan subspecies complex, *E. t. grabowskyi* and *E. t. callicyanous*, respectively. Neither native to China.

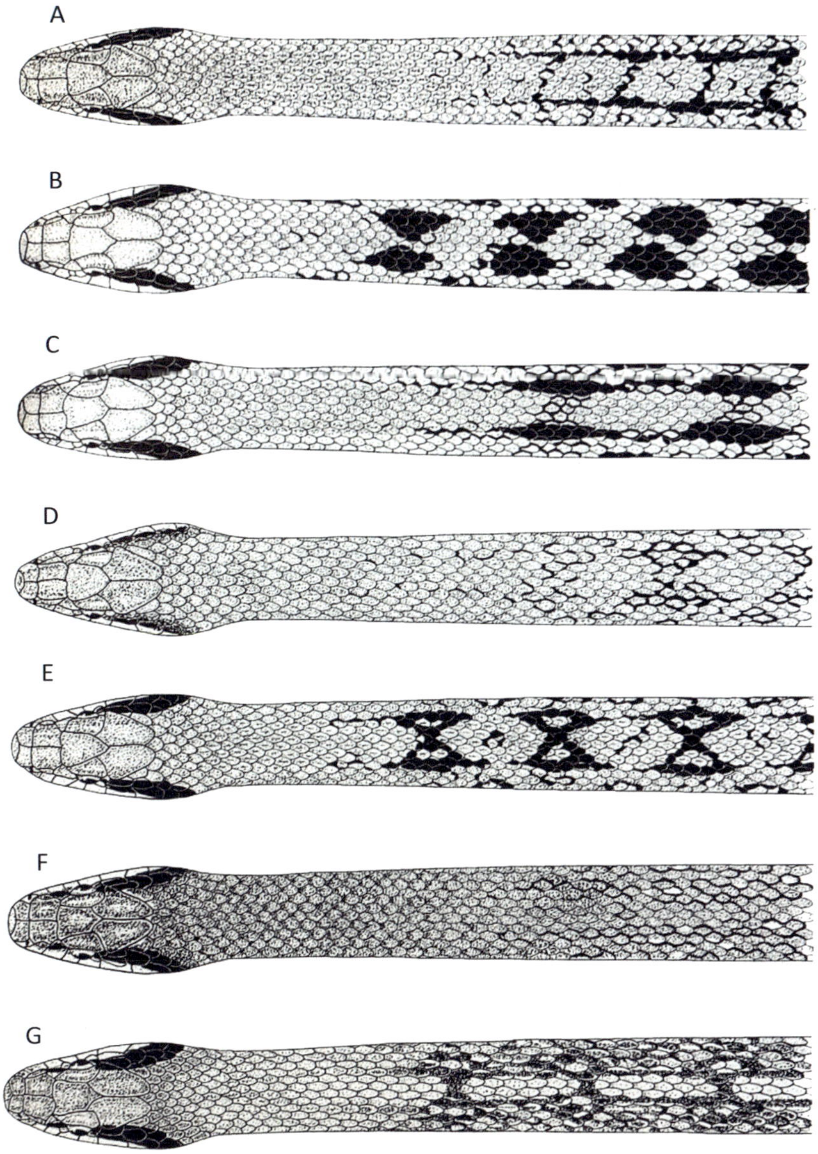

From Schulz (1996). A) *E. t. taeniura,* B) *E. t. friesi,* C) *E. t. mocquardi,* D) *E. t. schmackeri,* E) *E. t. yunnanensis,* F) *E. t. grabowskyi,* G) *E. t. callicyanous*

Left: *E. t. yunnanensis* from Kunming, Yunnan. Photo by Gernot Vogel. Right: *E. t. friesi* from northern Taiwan.

分类与命名：Schultz（1996，2010）将黑眉锦蛇所有亚种分为北（Indochinese subspecies）、南（Indomalayan subspecies）两个地域形式。北部地域形式经鉴定具以下五个亚种（A～E）。

A）*E. t. taeniura*：分布于中国东部。上唇鳞8（4-2-2、3-2-3、3-3-2式）或9（4-2-3、3-2-4、3-3-3式）；眶下鳞1；体中段背鳞23～25行；腹鳞225～255；尾下鳞84～112对。

B）*E. t. friesi*：分布于中国台湾。在冲绳岛上也有分布，不过为外来入侵物种。眶下鳞1或2；体中段背鳞25行；腹鳞243～262；尾下鳞96～123对。

C）*E. t. mocquardi*：分布于中国以及越南。上唇鳞9（4-2-3式）；眶下鳞1；体中段背鳞23～25行；腹鳞251～264；尾下鳞90～125对。

D）*E. t. schmackeri*：分布于琉球群岛。眶下鳞1；体中段背鳞25行；腹鳞246～260；尾下鳞104～125对。

E）*E. t. yunnanensis*：分布于不丹、中国、印度、缅甸、泰国和越南。近期，分布于北京的种群经鉴定为此亚种。上唇鳞7（3-2-2、4-2-1式）、8（3-2-3、4-2-2式）或9（3-2-4、4-2-3式）；眶下鳞无或1，少数2；体中段背鳞23～25行；腹鳞236～260行；尾下鳞89～120对。

F）& G）：*E. t. grabowskyi* 和 *E. t. callicyanous* 皆为南部地域形式下亚种，两者在中国皆无分布。

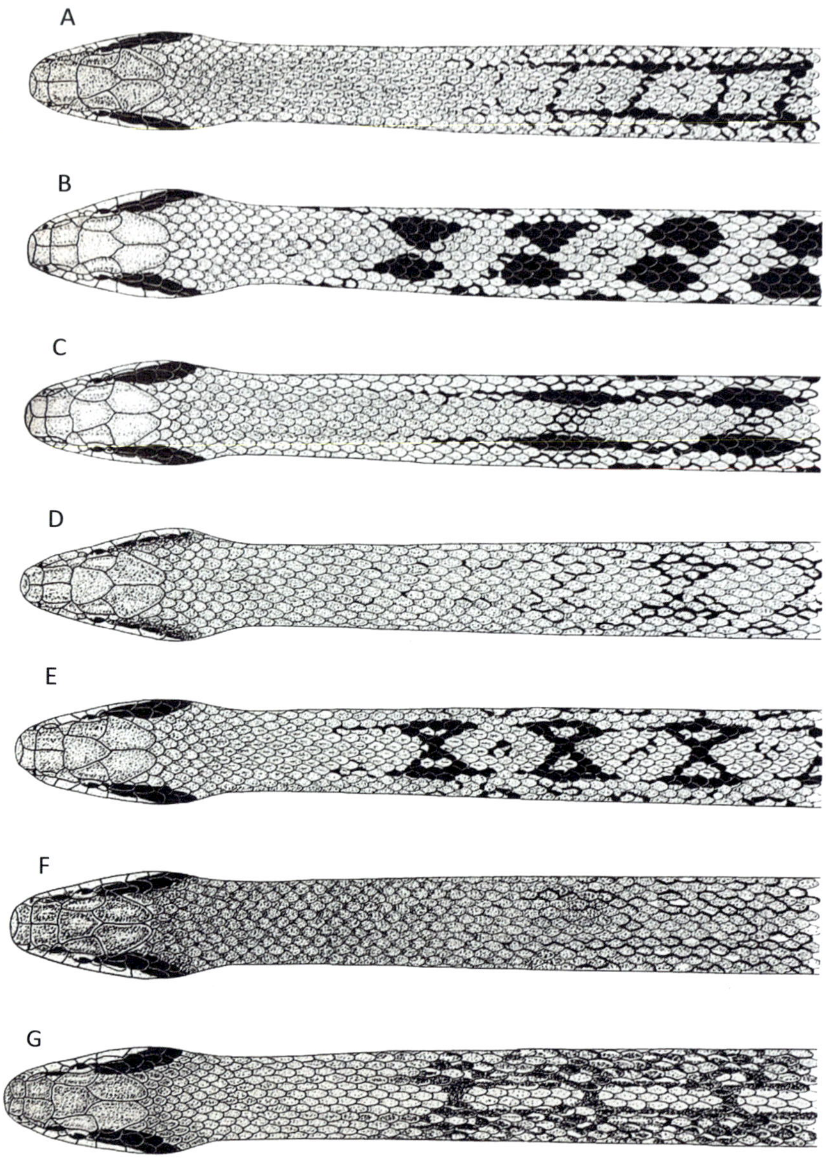

From Schulz (1996). A) *E. t. taeniura,* B) *E. t. friesi,* C) *E. t. mocquardi,* D) *E. t. schmackeri,* E) *E. t. yunnanensis,* F) *E. t. grabowskyi,* G) *E. t. callicyanous*

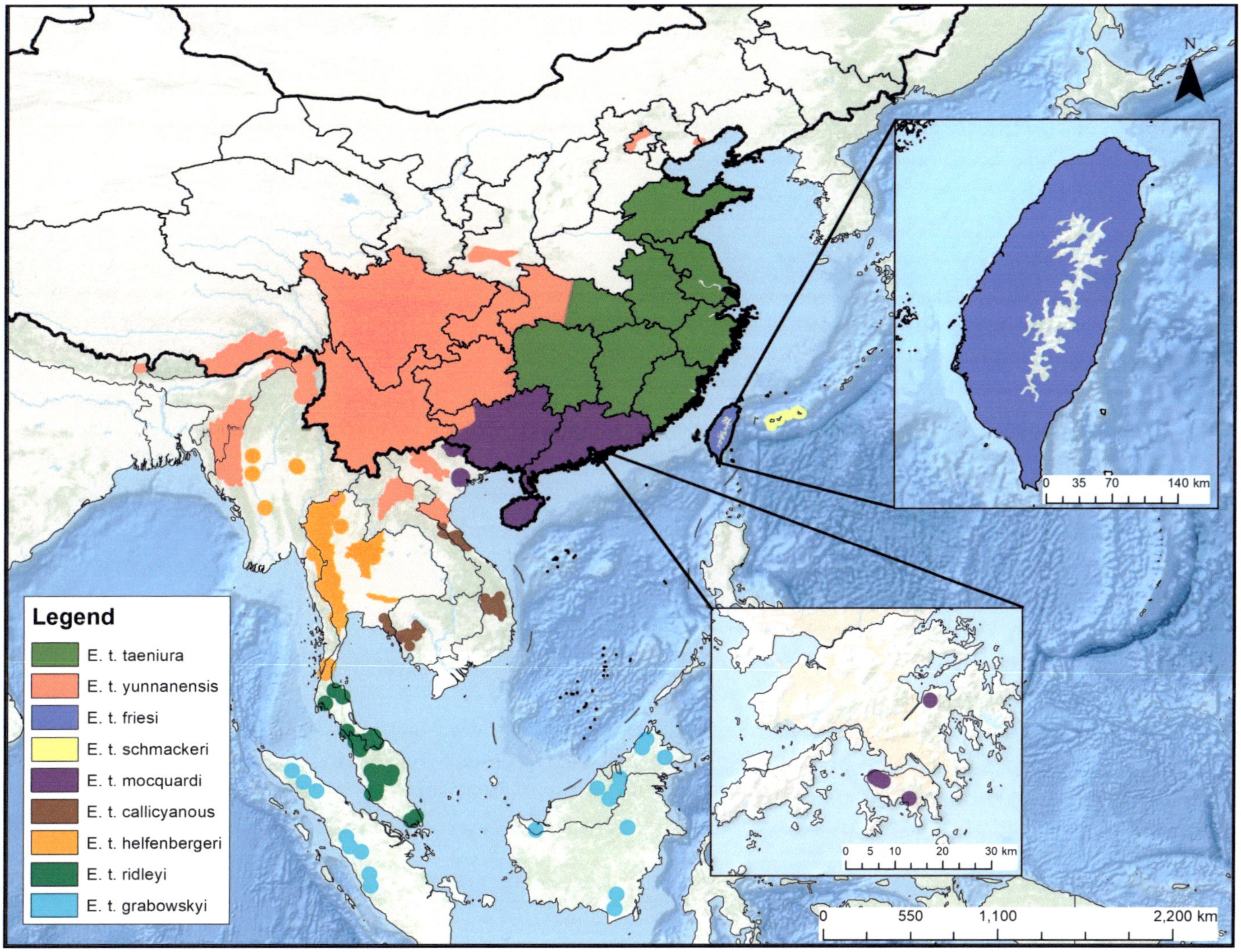

Precise distribution limits of supposed subspecies designations are estimations only, until further, more comprehensive studies have been conducted.

各亚种具体的地理分布仅基于推测，此方面更准确的信息仍待未来进一步的科学研究。

Key to the subspecies of *Elaphe taeniura* (Schultz, 2010). VS = width of vertebral stripe in dorsal scale rows, SC = subcaudals:

1A. Body compressed; 225-262 ventrals ········ 2
1B. Body very distinctly compressed; 272-305 ventrals ········ 3
2A. Dorsal pattern H-shaped, ladder-like interconnected, or fragmented ········ 4
2B. Dorsal pattern butterfly-shaped, not stretched; VS 4; neck without pattern for 1-3 head lengths ········ 5
2C. Dorsal pattern butterfly-shaped, stretched; VS 4; neck without pattern for 3-4 head lengths ········ *mocquardi*
3A. Ground color ochre to light grey; head bluish or grey; ventrals monochrome ········ 6
3B. Ground color yellowish, grey or cyan; head matching body in color ········ 7
4A. Snout 3-3.5 times the eye diameter; VS 3 ········ *schmackeri*
4B. Snout 2.5 times the eye diameter; VS 5 ········ *taeniura*
5A. Large oval spots; neck without pattern for 1-2 head lengths; tongue bluish-black ········ *friesi*
5B. Stretched spots often with light center; neck without pattern for 2-3 head lengths; tongue reddish-brown to black ········ *yunnanensis*
6A. No dorsal spots; lateral stripes without bands across; SC 105-122; anterior dorsal scales 29 ········ *ridleyi*
6B. Dorsal spots indistinct; lateral stripes segmented by light bands across; SC 82-97; anterior dorsal scales 27 ········ *helfenbergeri*
7A. Dorsal markings butterfly-shaped, stretched; lateral stripes light in color; VS 3; SC 116-125 ········ *callicyanous*
7B. Dorsal pattern indistinct, usually reticulated; VS 2; SC 92-114 ········ *grabowskyi*

黑眉锦蛇亚种检索（Schultz，2010）

1A. 体侧扁。腹鳞225～262 ········ 2
1B. 体侧扁十分明显。腹鳞272～305 ········ 3
2A. 体背具“H”形斑纹，呈梯状相连或互相断开 ········ 4
2B. 体背具蝴蝶状斑纹（展翅姿态正常）。体背中线处纵纹约占4行鳞宽。颈部1～3倍头长距离内无斑纹 ········ 5
2C. 体背具蝴蝶状斑纹（展翅姿态纵向拉伸）。体背中线处纵纹约占4行鳞宽。颈部3～4倍头长距离内无斑纹 ········ *mocquardi*
3A. 体背底色黄褐或浅灰色。头部淡蓝或灰色。腹面呈单一色 ········ 6
3B. 体背底色浅黄、灰或青色。头部、躯体同色 ········ 7
4A. 吻长是眼直径长度的3～3.5倍。体背中线处纵纹约占3行鳞宽 ········ *schmackeri*
4B. 吻长是眼直径长度的2.5倍。体背中线处纵纹约占5行鳞宽 ········ *taeniura*
5A. 蝴蝶状斑纹的“翅膀”部分呈椭圆形。颈部1～2倍头长距离内无斑纹。舌乌青色 ········ *friesi*
5B. 蝴蝶状斑纹的“翅膀”部分中心淡色。颈部2～3倍头长距离内无斑纹。舌红褐或黑色 ········ *yunnanensis*
6A. 体背无斑块。体侧具纵纹但无横斑。尾下鳞105～122对。颈部背鳞29行 ········ *ridleyi*
6B. 体背斑块不明显。体侧纵纹被浅色横斑切断。尾下鳞82～97对；颈部背鳞27行 ········ *helfenbergeri*
7A. 体背具蝴蝶状斑纹（展翅姿态纵向拉伸）。体侧纵纹色淡。体背中线处纵纹约占3行鳞宽。尾下鳞116～125对 ········ *callicyanous*
7B. 体背斑纹不明显，通常呈网状。体背中线处纵纹约占2行鳞宽。尾下鳞92～114对 ········ *grabowskyi*

Conservation Status: Not assessed by the IUCN.

保护状况：本种在《濒危物种红色名录》中被定为“未予评估”等级。

Comments: Based on ventral counts and body form, Schultz (1996, 2010) split the subspecies-complex into two geographic forms, a northern (the Indochinese subspecies) and southern (Indomalayan subspecies). The Indochinese complex has lower ventral scale counts, 225 to 264, whereas the Indomalayan complex has higher ventral scale counts, 272 to 305.

Definitive records from Russia and Korea are not known. A specimen in the Smithsonian Institute (USNM 14608) is supposedly from Korea, but exact locality data is not provided.

Regarding the population on Hong Kong. For many years it was curiously absent from Hong Kong. The first observation turned up in 2002. Father Anthony Bogadek, considered by many to be the most dedicated herpetologist in recent decades in Hong Kong, spent his entire life in the western district of Hong Kong Island (where the first specimens were found), and in all the years prior to 2002, a specimen was never turned up. As this species is very common in both the pet trade and as food or medicine, specimens are suspected to be intended releases, possibly from Buddhist monks practicing “mercy releases.” This is when monks purchase captive animals (regardless of the end fate) and release the animals in “natural areas,” under the misguided belief that they are helping the animals.

注释：根据腹鳞数目和体型，Schultz（1996，2010）将黑眉锦蛇所有亚种分为北（Indochinese subspecies）、南（Indomalayan subspecies）两个地域形式。北部诸亚种腹鳞数目较小，为225～264枚；南部诸亚种腹鳞数目较大，为272～305枚。

本种于俄罗斯及朝鲜半岛的分布情况尚不明确。据说美国史密森学会保有一枚采自于朝鲜半岛的标本（USNM 14608），不过具体采集地点不详。

关于本种于香港的分布，此前长时间未有目击记录。首次被人发现是在2002年：香港著名两栖和爬行动物学家鲍嘉天神父（Anthony Bogadek）在香港岛西部采集到了本种的首枚标本。考虑到黑眉锦蛇为宠物市场的常见物种，且常作食用及药用，故推测此次所发现的是人为放生个体。

Above: The Japanese subspecies, *E. t. schmackeri*, specimen from Okinawa Island, Japan (Gernot Vogel)

Below: The Taiwan subspecies, *E. t. friesi*, specimen from northern Taiwan.

The genus *Euprepiophis* / 玉斑蛇属

Etymology: *Euprepes* is Greek for "good looking, fair." *ophios* is Greek for "snake."

属名词源：源自希腊语，“*euprepes*”意为“美丽的”，“*ophios*”意为“蛇”（故该属有时也被称作“丽蛇属”）。

General captive care / 人工饲养概述：

If available, captive-bred specimens should be acquired. Wild-caught or farm-raised specimens typically have parasites, are malnourished, and are heavily stressed, resulting in a weakened immune system. If working with wild-caught or farm-raised individuals, a qualified veterinarian should conduct a fecal and lung wash to test for parasites. Animals should be quarantined according to the protocols set out by the facility or the keeper, but an approximate quarantine period of 3 months to 1 year is advised to prevent the spread of any potential diseases or parasites.

野外捕获个体及部分养殖场所提供的个体或因寄生虫感染、营养不良、应激反应等因素存在免疫力低下的隐患。若购入以上个体，建议联系有资质的兽医对个体的呼吸道和消化道样本进行寄生虫检测。若可能，应从可靠渠道获取人工繁殖个体。新入手的个体需实行隔离，隔离时长可根据提供方给出的建议。一般建议隔离3个月至1年。

This genus is fossorial and should be treated as such in captivity with deep layers of leaf litter and/or mulch for the individuals to burrow under. *E. mandarinus* can often be found buried in the substrate with a temperature range of 21-25°C and relative humidity of 60-70%. Adult and juvenile specimens should be offered a minimum terrarium of 86 x 46 x 16 cm. Males of *E. mandarinus* should not be housed together as they will engage in combat during the breeding season, often resulting in one of the individuals' death. In nature, the inferior male would leave the territory after being defeated by the dominant male. When males are housed together in a confined space, this does not give the inferior male a chance to leave the dominant male's territory. Hatchlings can be housed in a plastic shoebox or commercialized snake breeding containers measuring 31 x 19 x 10 cm for the first 1-2 years of life, after which they should be transferred to a larger enclosure. Paper towels are used as the substrate for hatchlings, as this genus has a very strong feeding response and is used to avoid accidental ingestion of substrate debris. Hatchling setups should also include a moist hide which contains sphagnum moss, which is commonly used when the juveniles are opaque. Some neonates may not start feeding immediately after hatching and should be offered a short cooling period to stimulate the endocrine system.

A high yellow specimen (Matthew Most).

After a brief period of 2-3 weeks at similar temperatures as those during brumation, these individuals will usually start feeding. As with other species, these species are also cycled with weather patterns. These species are maintained at temperatures of 10-15°C for 3-4 months. To induce breeding behavior, males should be introduced to females following brumation.

人工饲养下，本属的蛇种有钻入底材的习性，故底材的铺设应达到一定的厚度。底材可选用落叶、树皮或木屑。当环境温度达到21～25°C，相对湿度介于60～70%，玉斑锦蛇（*Euprepiophis mandarinus*）或钻入底材之中。成体及小亚成体的饲养容器应不小于86 x 46 x 16厘米。雄性玉斑锦蛇在发情期会互相打斗，常导致伤亡，故切不可群养。幼体可饲养在普通塑料盒或专业饲养箱内，空间大小建议在31 x 19 x 10厘米。不过随着幼体的成长，1～2年后应考虑为其提供更大的饲养空间。考虑到本属蛇种有较强的摄食反应，为避免误食底材，幼体的底材建议选用厨房纸。幼体的环境中应放置加湿躲避穴，可在其中填充泥炭藓用于提高湿度。小亚成体蜕皮时也常使用该类躲避穴。部分刚孵化个体在出壳后或不会立即进食，此时应短暂降温以刺激其内分泌系统。2～3周的低温期后（温度与冬化时的温度一致），个体往往会开始进食。为模拟本属蛇种在自然环境中的节律，个体应经历为期3～4个月的冬化，冬化时的温度应保持在10～15°C。为诱发繁殖行为，公母个体应在冬化后置于同一饲养环境中。

Egg Incubation: Eggs of this genus are housed in commercially available incubators set to 25-27°C or room temperature and measuring a minimum of 20 x 20 x 15 cm. Incubating at room temperature will result in a hatching delay by approximately 10-15 days. The containers must be well ventilated with holes on all sides to allow for fresh air flow and remove excessive moisture. Excessive moisture and improper ventilation can create issues for developing embryos/eggs, as mold and fungus can grow on the egg's surface. Eggs can be incubated upon a variety of media, but preference is vermiculite and perlite. Both substrates retain moisture and aid in maintaining humidity. Still, both media can dry out and may require additional water if necessary. To achieve proper moisture for incubating, dry media should be measured to have a 5 cm depth within the container. A 2:1 ratio of substrate to water by weight is normally sufficient to ensure proper moisture and humidity during incubation. One can check the appropriate balance by squeezing the mixture by hand. Expressing water means the water content is too high. After achieving the proper moisture, the media, followed by the eggs, can be placed in the incubator. When eggs are laid, it is best to gently separate eggs to allow for adequate airflow within the incubation chamber. Eggs are placed atop the substrate or partially sunk into the media to hinder the egg's movement when other eggs are hatching.

A freshly hatched specimen (Matthew Most).

卵的孵化：使用专业孵化器以25～27℃或室温进行孵化皆可使本属蛇种的幼体顺利出壳，但室温下的孵化期要多10～15天。孵化器的尺寸建议不小于20 x 20 x 15厘米。盛放卵的孵化盒四周均应有孔，以保证空气的流通及多余湿气的排出。湿度过高或通风不畅将对卵的发育造成影响，易令卵壳表面滋生真菌或霉菌。孵化介质有多种选择，推荐使用蛭石或珍珠岩。两者皆有不俗的锁水性，有利于湿度的保持。然而，两种介质也可能在孵化的过程中变得过干。若有必要，需另行补水。为确保孵化过程中湿度始终处于理想水平，孵化介质（干燥状态下）在容器内的铺设厚度需达到5厘米。然后，水和孵化介质在重量上以1:2进行配比。一种检验配比是否合适的有效方法是用手捏兑水后的孵化介质。若有水渗出，则表明水份过高。放置卵时，建议互相之间保持一定的距离，以促进孵化盒内空气的流通。卵可放置在孵化介质表面或半埋于孵化介质之中。若采取半埋的方式则可限制卵的移动，从而避免对孵化造成影响。一切准备妥帖之后，孵化盒放入孵化器中。

Euprepiophis mandarinus specimen from northern Taiwan (Kevin Caldwell).

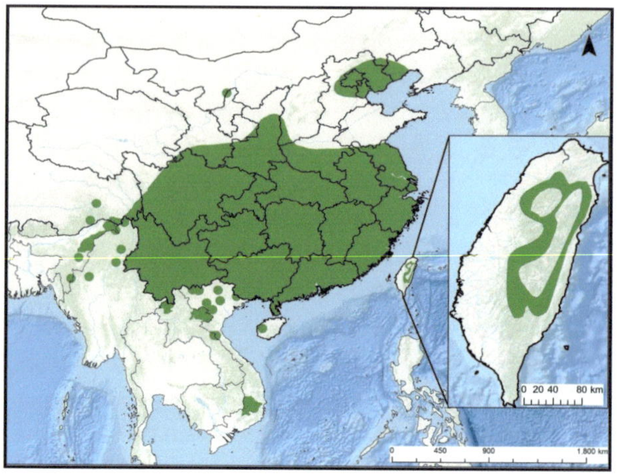

Scientific name / 学名: *Euprepiophis* [*Elaphe*] *mandarinus* (Cantor, 1842)

English name / 英文名: Mandarin ratsnake, Mandarin trinket snake

Chinese name / 中文名: 玉斑锦蛇 / yù bān jǐn shé

Etymology / 种名词源: *Mandarinus* is a Latinized form of the word "Mandarin." / "*mandarinus*" 为"mandarin"一词的拉丁化形式，该词在此意指"中国"。

Terra typica / 模式产地: Zhoushan Island, Zhejiang, China / 中国，浙江省，舟山岛

Length / 体长: Maturity at 750-850 mm; Hatchlings emerge around 180-380 mm. Hatchlings from Vietnam measure around 260-380 mm. The maximum size recorded is 1740 mm / 体长750～850毫米时达到性成熟；刚孵化的幼蛇体长180～380毫米（分布于越南的玉斑锦蛇幼体出壳时体长260～380毫米）。已知最大个体体长1740毫米。

Description: This species is a medium-sized species. The cephalic scales are as follows: supralabials 6-8, 3rd and 4th, or 4th and 5th in contact with orbit; infralabials 8-10; loreal 1; preocular 1, large; postoculars 2; temporals 2+2. Dorsal scales in rows of 17-19: 19-23: 17-19. The scales are smooth and shiny. Ventrals 210-240; subcaudals 62-80, paired, with a divided anal plate. The loreal scale may be small, absent, or partially or completely fused with the prefrontal. The head is oval with a marginally blunt snout; the tail is medium in length. The eye is medium-sized with a dark brown to black iris with a round pupil, which is less clear from the rest of the eye. The head is made up of an intricate pattern of black and yellow cross-bands, sometimes forming faint spear-points towards the posterior region of the head. The labials are white, except for where the black cross-bands from the top of the head intersect with the labials. The upper body and tail are purplish-gray or even reddish. The dorsum consists of a series of black diamonds with a yellow center and a thin yellow margin in typical specimens. In some specimens, there may also be a series of small lateral black spots. The underside of the head is white except for black spots on the gulars and some infralabials. The underside of the body and tail is white with scattered quadrangular marks of prominent black pigment, unevenly arranged on sides. Juveniles resemble adults but are usually more vibrantly colored.

Specimens from Shennongjia, in western Hubei.

外观描述：体型中等。鳞被：上唇鳞6（2-2-2、3-2-1式），7（2-2-3、3-2-2式）或8（2-2-4、3-2-3式）；下唇鳞8～10；颊鳞1；眶前鳞较大，1枚；眶后鳞2；颞鳞2+2；背鳞平滑亮泽，颈部17～19行，中段19～23行，肛前17～19行；腹鳞210～240；尾下鳞62～80对；肛鳞二分；颊鳞变数较大，或较小，或缺失，或与前额鳞合一。头椭圆形，吻略钝。尾长短适中。眼大小适中，瞳孔圆形，虹膜深褐或黑色。头背黄色，具3条黑斑：第一条横跨吻背；第二条横跨于双眼；第三条位于头后方，呈"V"字形。躯尾背面紫灰或淡红色，中央具一列黑色菱形斑块。斑块中心为黄色，外侧亦镶以黄薄边。个别体两侧各有一排黑色小斑点。头腹面整体白色，喉及唇部散有黑斑。躯尾腹面白色，两侧散有交错的黑色方斑。小亚成体在外观上与成体相似，不过体色通常更为艳丽。

Distribution: This species has a very wide range within Greater China, being found in Anhui, Beijing, Chongqing, Fujian, Gansu, Guangdong, Guizhou, Hebei, Henan, Hubei, Hunan, Jiangsu, Jiangxi, Liaoning, Ningxia, Shaanxi, Shanghai, Shanxi, Sichuan, Taiwan, Tianjin, Tibet, Yunnan, and Zhejiang. Seemingly absent from Hainan Island, though there are occasional sightings, these are believed to be escaped pets. There have not been any confirmed breeding or wild populations found on Hainan Island. The species is also absent from Hong Kong. Outside of Greater China, this species is found in India, Laos, Myanmar, and Vietnam. Found from 450 to 3000 m.

地理分布：广布种。在中国，分布于安徽、北京、重庆、福建、甘肃、广东、贵州、河北、河南、湖北、湖南、江苏、江西、辽宁、宁夏、陕西、上海、山西、四川、台湾、天津、西藏、云南和浙江。

Specimen from Vietnam (Matthew Most).

A large, road-killed specimen from northern Guangdong.

海南似乎未有分布，虽偶有目击记录，但推测为脱逃至野外的人工饲养个体。至今，尚未有官方信息确认海南岛有野生种群存在。香港也未有分布。中国以外，分布于印度、老挝、缅甸和越南。垂直分布450～3000米。

Habits and habitats: *E. mandarinus* are known to inhabit montane forests, forested woodlands, environments with scattered rocks and bushes, and even disturbed areas such as agricultural areas and human dwellings. They are frequently found in higher altitudes and during cooler temperatures than many other ratsnake species. They are a very secretive snake that spends most of its time in burrows searching for its favorite prey of young rodents. They are primarily crepuscular and nocturnal, being active in the early morning, late afternoon, or night. When threatened, they will often vibrate their tail, strike repeatedly, and not hesitate to bite and musk if picked up.

习性与栖息地：已知栖息于山林、森林、散有碎石和灌木的环境，也见于农田和人类住所附近。相较于其它以"锦蛇"为名的蛇种，玉斑锦蛇经常在海拔更高、温度更低的环境中被人发现。本种生性隐秘，大部分时间都在洞穴中寻觅其最钟爱的食物—啮齿类幼崽。主要在黎明、黄昏或夜间活动。若感知威胁，常抖动尾部并反复发动攻击。若被抓取会立即扑咬并分泌恶臭液体。

Diet and feeding habits: In captivity, the species will take small mammals such as rodents. In the wild, this species likely preys on rodents' nests, as juvenile and adult animals in captivity prefer multiple smaller rodent prey items. Captive hatchlings feed on newborn mice readily without scenting. Prey is killed via constriction.

食性与喂养：人工饲养个体主要以小型哺乳动物如啮齿类为食。人工繁殖的幼体可直接进食在气味上未经特殊处理的乳鼠。根据人工饲养下的食性，推测本种在自然环境中捕食啮齿类刚出生的幼崽。有绞杀猎物的习性。

Typical examples of habitat of *E. mandarinus* in Shennongjia, Hubei, China.

Captive care: Adult and juvenile specimens prefer a deep layer of mulch and leaf litter as a substrate. In captivity, this species has been recorded to have a life span of 10-15 years.

人工饲养：成体和小亚成体饲养环境中底材的铺设应达到一定的厚度。底材可选用落叶、树皮或木屑。据记载，人工饲养下本种的寿命为10～15年。

Reproduction: Oviparous. *E. mandarinus* reach sexual maturity in three to four years. In captivity, specimens are provided with a brumation period of 10-15℃ for 3-4 months. Copulation is observed in both the spring and summer months. Chinese locality animals will produce two clutches in a season, with eggs being laid in May-June and potentially a second clutch in September. Vietnamese locality animals will only lay one clutch a year and typically breed in the late spring, April-May. The gestation period for *E. mandarinus* is 50-60 days with clutch size ranging from 2-10 eggs, measuring 50-57 x 23-27 mm, and requiring incubation of eggs at 23-25℃ for 60-70 days. Males may subdue the female by biting on the back of the neck of the prospective mate. If some hatchlings will not eat, they should be offered a short cooling period to stimulate the endocrine system. After a brief period of 2-3 weeks at similar temperatures as those during brumation, these individuals will usually start feeding. Most hatchlings and juveniles prefer multiple smaller prey items and likely raid rodent nests in the wild.

繁殖：卵生。3～4年达到性成熟。人工饲养下，个体应在10～15℃下经历3～4个月的冬化期。据观察，本种在春季和夏季皆有交配行为。分布于中国的个体一年产卵两窝：一窝产于5～6月，另一窝则可能在9月。分布于越南的个体一年只产卵一窝，交配则通常在晚春4～5月进行。母蛇怀卵50～60天。一窝卵2～10枚。卵长径50～57毫米，短径23～27毫米。孵化温度23～25℃，孵化期60～70天。交配时，雄性或通过咬住雌性颈背来制服雌性。部分刚孵化个体在出壳后或不会立即进食，此时应短暂降温以刺激其内分泌系统。在2～3周的低温期后（温度与冬化时的温度一致），个体往往会开始进食。大部分幼体及小亚成体偏爱体型较小的食物，故推测其在野外捕食啮齿类的幼崽。

E. mandarinus hatching from eggs (Matthew Most).

In captivity, this species can be extremely variable and arguably among the most beautiful Asian snakes. Many color mutations have been established in the reptile trade, including axanthic animals from Hong Valley Sapa, Vietnam, which hatch out a bright yellow, which then fades to white with age. Some specimens also express xanthochromism, which brings about a very intense yellow.

玉斑锦蛇人工种群的个体在外观上极为多变，可谓亚洲最美丽的蛇种之一。目前，宠物市场已培育出多个色系的变异个体，包括一种来自于越南老街省沙坝县的无黄色素个体。该变异个体在刚孵化出壳时呈亮黄色，但体色会随着成长逐步褪为白色。同时，也存在着高黄体色的个体。

A variety of colorations and morphs known in captivity (all specimens by Matthew Most).

Taxonomy and nomenclature: There are no known subspecies of this wide-ranging species.

分类与命名：本广布种目前无已知的亚种。

Similar species: It would be very difficult to confuse this species with any other species, aside from perhaps the extremely rare *Euprepiophis perlaceus*. Outside of areas of sympatry, this species should be easy to distinguish from any other species.

相似物种：除或同极罕见的横斑锦蛇（*Euprepiophis perlaceus*）相混淆外，玉斑锦蛇极易同其它蛇种相区分。

Above: *E. mandarinus* from western Hubei. Below: *E. perlaceus* from western Sichuan (Ding Li 丁利).

Conservation status: Considered Least Concern (LC) according to the IUCN Redlist, last assessed in 2011.

保护状况：本种在《濒危物种红色名录》中被定为“无危”等级（最近一次评定为2011年）。

Typical habitat of *E. mandarinus* from Wuyishan, Fujian, China.

Comments: Three main color background variants. A reddish background, a grayish background, and yellow background. In Taiwan, this species is known locally by a different Chinese name, it is known as: 高砂蛇 gāo shā shé.

注释：体背底色主要有三种颜色的变异：分别为淡红、浅灰和黄色。在中国台湾，本种被称为“高砂蛇”。

A-B: Specimens from Wuyishan, northern Fujian. C-D: Specimens from Shennongjia, western Hubei. E: Specimen from Zhejiang (Matthew Most). F: Specimen from Baling, Taiwan (Kevin Caldwell).

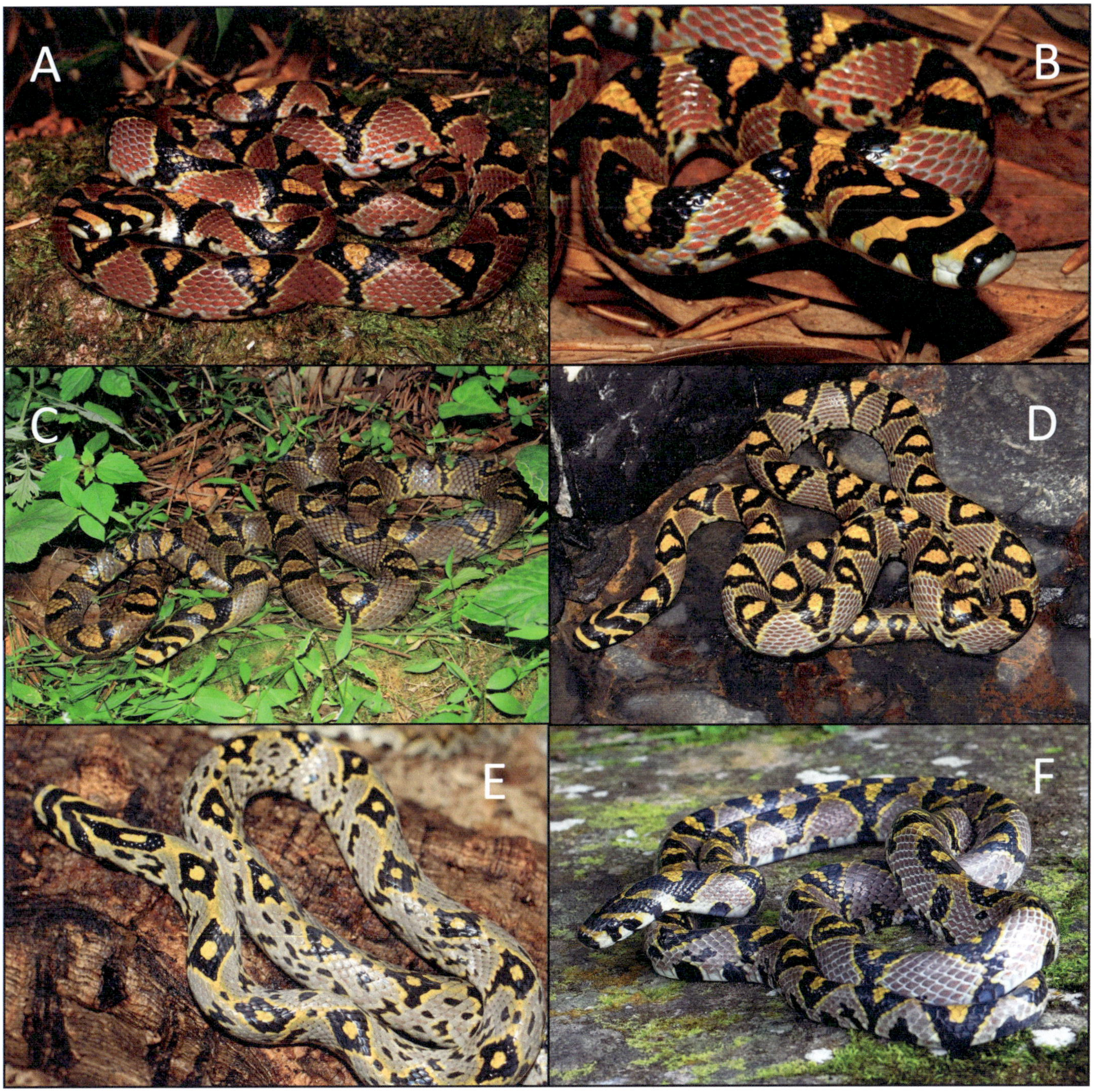

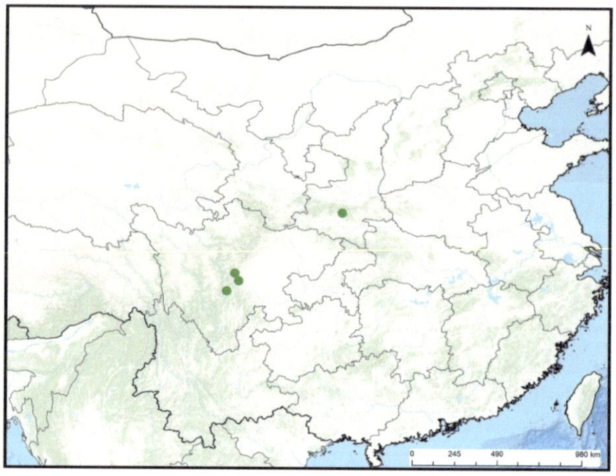

Scientific name / 学名: *Euprepiophis* [formerly *Elaphe*] *perlaceus* (Stejneger, 1927)

English name / 英文名: Sichuan ratsnake, Pearl Banded ratsnake

Chinese name / 中文名: 横斑锦蛇 / héng bān jǐn shé

Etymology / 种名词源: *Perlaceus* is Latin for "pearls," in reference to the "pearl bands" of this beautiful species / "*perlaceus*"，拉丁语，意为"珍珠"。该种名意指本种体背珍珠状的横斑。

Terra typica / 模式产地: Ya'an, Sichuan, China / 中国，四川省，雅安市

Length / 体长: Maturity at 750-850 mm; Hatchlings emerge around 180-360 mm / 体长750～850毫米时达到性成熟；刚孵化的幼蛇体长180～360毫米。

Description: The cephalic scales are as follows: supralabials 7, 3rd and 4th in contact with orbit; infralabials 7; loreal present; preocular 1; postocular 2; temporals 1+2; dorsal scales in 19:19:17 rows with the dorsal most scales feebly keeled; ventral scales 226-229, obtusely angulate laterally; anal divided; 69 pairs of subcaudals. There is a slight ontogenetic coloration and pattern shift from juvenile to adult. Juveniles are tawny grey with several dorsal yellow bands, bordered by black on either side, stopping at the ventral scales. Not all bands may be complete. The venter consists of black and white checkered squares irregularly spaced throughout. Dorsal body bands may number 34 to 37, with 10-11 bands on the tail. The pattern on the head is similar to *E. mandarinus*, which this species is closely aligned with. The head consists of a series of black and yellow crossbands with an approximate spear-point shaped pattern converging on the frontal scale.

外观描述：鳞被：上唇鳞7（2-2-3式）；下唇鳞7；具颊鳞；眶前鳞1；眶后鳞2；颞鳞1+2；背鳞19-19-17行，多数背鳞微弱起棱；腹鳞226～229，侧棱不明显；肛鳞二分；尾下鳞69对。小亚成体与成体在体色和斑纹上略有差异。小亚成体背面灰褐色，具黄色横斑，横斑镶有黑边，止于腹面。个别横斑或不完整。腹面散有黑色方斑。体背横斑数量34～37个，尾部10～11个。头部斑纹与玉斑锦蛇头部的斑纹相似：一道黑斑横跨于两鼻孔间并延伸至唇部；另一道黑斑横跨于双眼并在眼下分叉延伸至唇部；头枕部有2个呈叠套的"V"形黑斑。

Adult specimens of *E. perlaceus* from Sichuan (left: Gernot Vogel; right: Ding Li 丁利)

Distribution: Sichuan and Shaanxi, China. Li and Rong (2001) recorded the species from Nanling Nature Reserve in Guangdong Province, China, which is incorrect. There are new records from Shimian and Baoxing Counties, Sichuan (Hu et al. 2002).

地理分布：分布于中国的四川和陕西。多年前，四川省雅安市下的石棉县和宝兴县也发现有本种分布（胡杰等人，2002）。本种于广东省南岭国家级自然保护区的分布记录并不正确。

Habits and habitat: Found in high valleys from 1600—2400 m. The vegetation type is moist mountain deciduous broadleaf forest, as well as shrub and farmlands. The habits of this species are likely very similar to *E. mandarinus*. See the species *E. mandarinus.*

习性与栖息地：见于海拔1600～2400米的山谷之中。栖息地的植被类型包括潮湿的落叶阔叶林、灌木以及农田。本种与玉斑锦蛇近缘，两者在习性上或极为相似，具体情况可参阅玉斑锦蛇介绍章节中的“习性与栖息地”版块。

Diet and feeding habits: In captivity, the species will take small mammals such as rodents. In the wild, this species likely preys on rodents and bird nests as juvenile and adult animals in captivity prefer multiple smaller prey items.

食性与喂养：人工饲养个体主要以小型哺乳动物如啮齿类为食。根据人工饲养下的食性，推测本种在自然环境中捕食幼鸟和啮齿类幼崽。

Captive care: Captive care of this species is poorly known. Considering the similarities to *E. mandarinus,* recommendations for captive care would follow suit. Adult and juvenile specimens likely prefer a deep layer of mulch and leaf litter as substrate.

人工饲养：对此知之甚少。考虑到本种与玉斑锦蛇有诸多相似之处，具体饲养建议可参阅玉斑锦蛇介绍章节中的“人工饲养”版块。成体和小亚成体饲养环境中底材的铺设应达到一定的厚度。底材可选用落叶、树皮或木屑。

General habitat of *E. perlaceus* in Sichuan.

Reproduction: Oviparous. Sexual dimorphism is observed between males and females, with females being larger than males. *E. perlaceus* likely reaches sexual maturity in 3-4 years based upon similarities to *E. mandarinus*. A handful of keepers kept this species and reported breeding in captivity occurred in 2017 at the Chinese Academy of Sciences. Recommendations for captive breeding would follow *E. mandarinus* with a brumation period of 10-15℃ for 3-4 months. Copulation is likely in the spring, with eggs being laid in June-July.

繁殖：卵生。据观察，雌雄个体之间存在性别二态性，雌性体型大于雄性。考虑到与玉斑锦蛇有诸多相似之处，推测本种或于3～4年达到性成熟。人工繁殖方面也可参考玉斑锦蛇，建议个体在10～15℃下经历3～4个月的冬化期。交配或在春季进行。产卵或在6～7月。仅极少数机构和个人饲养本种。据报道，中国科学院在2017年成功实现横斑锦蛇的人工繁育。

Juvenile *E. perlaceus* possess more contrasting colors and have unbroken black bands which gradually develop yellow highlighted scales in adults, creating a more broken appearance. Photos above and below (Ding Li 丁利).

Taxonomy and nomenclature: There are no known subspecies. For many years, several people wondered if this species was just a pattern variation of *E. mandarinus*, but several follow-up genetic studies have confirmed this species' validity.

分类与命名：目前无已知亚种。多年来，一些人士猜测横斑锦蛇或仅为玉斑锦蛇的斑纹变异个体，但多个遗传研究已确认本种为有效物种。

Similar species: It would be very difficult to confuse this species with any other species, aside from perhaps the more common *Euprepiophis mandarinus*, which possesses yellow and black diamonds instead of cross-bands. Outside of areas of sympatry, this species should be easy to distinguish from any other species. To the casual observer, some people may confuse young vibrant specimens with *Lycodon flavozonatus* (pictured right).

相似物种：除或同较常见的玉斑锦蛇相混淆外，横斑锦蛇极易同其它蛇种相区分。玉斑锦蛇体背具黄黑菱形斑纹，与横斑锦蛇的斑纹有显著差别。非专业人士也或将年轻、体色艳丽的横斑锦蛇同黄链蛇（*Lycodon flavozonatus* / 右图）相混淆。

Conservation Status: Considered Endangered (EN) according to the IUCN Redlist, last assessed in 2011.

保护状况：本种在《濒危物种红色名录》中被定为"濒危"等级（最近一次评定为2011年）。

Specimen from Chengdu (Gernot Vogel).

The genus *Gonyosoma* / 树栖锦蛇属

Etymology: *Gonyosoma* = "Edged belly" or "venter with an edge," referring to the edge or keel along the side of the ventral scales, which assist in gripping rough surfaces to aid in climbing. Many arboreal species have such adaptations.

This genus is composed of relatively medium to large (80 to 240 cm) species, with a slender, laterally compressed body, that is generally green in coloration. Heads are elongate and distinct from the neck, with relatively large eyes. Tails are relatively long (20% to 30% of total length) and slender. Most members have an arboreal lifestyle.

属名词源："*Gonyosoma*"意为"有边的腹部"，意指腹面上的侧棱。侧棱有助于增加攀附时的摩擦力，许多树栖蛇种都有这一特征。

本属所辖皆为中大型蛇种（80～240厘米），身型修长，体侧扁，体色通常为绿色。头部较长，与颈部区分明显。眼部相对较大。尾部细长，约占体长的20～30%。大多数成员具有树栖习性。

General captive care / 人工饲养概述:

If available, captive-bred specimens should be acquired. Wild-caught or farm-raised specimens typically have parasites, are malnourished, and are heavily stressed, resulting in a weakened immune system. If working with wild-caught or farm-raised individuals, a qualified veterinarian should conduct a fecal and lung wash to test for parasites. Animals should be quarantined according to the protocols set out by the facility or the keeper, but an approximate quarantine period of 3 months to 1 year is advised to prevent the spread of any potential diseases or parasites.

野外捕获个体及部分养殖场所提供的个体或因寄生虫感染、营养不良、应激反应等因素存在免疫力低下的隐患。若购入以上个体，建议联系有资质的兽医对个体的呼吸道和消化道样本进行寄生虫检测。若可能，应从可靠渠道获取人工繁殖个体。新入手的个体需实行隔离，隔离时长可根据提供方给出的建议。一般建议隔离3个月至1年。

Juveniles and adults of this genus prefer an arboreal terrarium with a minimum size of 46 x 46 x 61 cm per adult specimen. Adult and juvenile specimens should be offered full-spectrum lighting as they are commonly observed in the branches above the ground in the mid-morning and early afternoon. Individuals prefer a temperature range of 21-25°C and relative humidity of 50-60%. It is common to see these species in their water bowl, so it is important to provide clean, fresh water regularly as they can develop fungal or bacterial infections in their interstitial skin if left in unsanitary conditions for too long. Hatchlings can be housed in a plastic shoebox or commercialized snake breeding containers measuring 31 x 19 x 10 cm for the first 1-2 years of life, after which they should be transferred to a larger enclosure.

本属蛇种成体和小亚成体的饲养环境布置应以树栖为主。单一成体的饲养容器应不小于46 x 46 x 61厘米。成体与小亚成体常可见在上午和午后盘于枝杈晒太阳，故饲养时应为其提供全光谱照明。人工饲养下，环境温度应达到21～25°C，相对湿度50～60%。据观察，本属蛇种经常盘踞于水盆之中，故保证水源的清洁至关重要，否则极易引起皮肤感染等问题。幼体可饲养在普通塑料盒或专业饲养箱内，空间大小建议在31 x 19 x 10厘米。不过随着幼体的成长，1～2年后应考虑为其提供更大的饲养空间。

Top and below images: *G. prasina:* this species should be given plenty of climbing opportunities (Matthew Most).

Paper towels are used as the substrate for hatchlings, while cypress mulch is used for adults. *G. frenatum* and *G. boulengeri* hatchlings prefer a shallow water bowl as they will be more apt to feed on live fish and rodents in their water bowl than on the dry surface. It has been hypothesized that these species have both a sensory nerve network on the front of the rostrum to detect movement and the neonate coloration serves as an adaptation to feeding in an aquatic setting as neonates. With weather patterns, these animals are cycled in the fall, with temperatures gradually cooling to stimulate brumation and follicular development in females. These animals are maintained at a temperature of 10-15°C for 3-4 months. They are slowly brought up to temperature in the spring and offered food before introducing males and females for breeding.

幼体的底材建议选用厨房纸，成体的底材可选用柏树皮。灰腹绿锦蛇（*Gonyosoma frenatum*）和尖喙蛇（*Gonyosoma boulengeri*）幼体的饲养环境中应放置浅水盘。相较于在干燥表面上，两者的幼体更偏爱在水盘中进食活鱼或啮齿类。据推测，灰腹绿锦蛇和尖喙蛇幼体的吻端具特殊的神经网络，可感知周围环境中的变化，有助于它们在水中觅食；同时，两者幼体特殊的体色起到保护色的作用，有助于它们躲避天敌。为模拟本属蛇种在自然环境中的节律，饲主应在秋季逐步降低饲养环境的温度以刺激个体的冬化行为以及雌性卵泡的发育。冬化时的温度应保持在10～15℃，为期3～4个月。冬化后，饲主应在春季缓慢升温并在喂食之后再将公母个体置于同一饲养环境中交配繁殖。

Egg Incubation: Eggs of this genus are housed in commercially available incubators set to 25-27°C or room temperature and measuring a minimum of 20 x 20 x 15 cm. Incubating at room temperature will result in a hatching delay by approximately 10-15 days. It is best to set these eggs at lower temperatures as the embryo can absorb the yolk sac fully. While the incubation period maybe longer, the neonates are larger and accept larger prey items. The containers must be well ventilated with holes on all sides to allow for fresh airflow and remove excessive moisture. Excessive moisture and improper ventilation can create issues for developing embryos/eggs, as mold and fungus can grow on the egg's surface. Eggs can be incubated upon a variety of media, but preference is vermiculite and perlite. Both substrates retain moisture and aid in maintaining humidity. Still, both media can dry out and may require additional water if necessary. To achieve proper moisture for incubating, dry media should be measured to have a 5 cm depth within the container. A 2:1 ratio of substrate to water by weight is normally sufficient to ensure proper moisture and humidity during incubation. One can check the appropriate balance by squeezing the mixture by hand.

One way of suspending eggs above vermiculite substrate (Ryan Goodman and Benjamin Levin, R & B Reptiles).

Expressing water means the water content is too high. After achieving the proper moisture, the media, followed by the eggs, can be placed in the incubator. When eggs are laid, it is best to gently separate eggs to allow for adequate airflow within the incubation chamber. Eggs are placed atop the media, and a means of suspending the eggs over the media is used for this genus to avoid undue moisture uptake. As a precaution, eggs should not be buried in the substrate for this genus.

卵的孵化：使用专业孵化器以25～27℃或室温进行孵化皆可使本属蛇种的幼体顺利出壳，但室温下的孵化期要多10～15天。孵化器的尺寸建议不小于20 x 20 x 15厘米。此处推荐以低温进行孵化，这有助于胚胎对卵黄囊充分吸收。孵化期将会延长，但出壳后的幼蛇体型明显更大，这也有利于其进食更大的食物。盛放卵的孵化盒四周均应有孔，以保证空气的流通及多余湿气的排出。湿度过高或通风不畅将对卵的发育造成影响，易令卵壳表面滋生真菌或霉菌。孵化介质有多种选择，推荐使用蛭石或珍珠岩。两者皆有不俗的锁水性，有利于湿度的保持。然而，两种介质也可能在孵化的过程中变得过干。若有必要，需另行补水。为确保孵化过程中湿度始终处于理想水平，孵化介质（干燥状态下）在容器内的铺设厚度需达到5厘米。然后，水和孵化介质在重量上以1:2进行配比。一种检验配比是否合适的有效方法是用手捏兑水后的孵化介质。若有水渗出，则表明水份过高。放置卵时，建议互相之间保持一定的距离，以促进孵化盒内空气的流通。卵可放置在孵化介质表面或悬于孵化介质之上。第二种方法可有效降低因湿度过高对孵化造成的不利影响。切不可将本属蛇种的卵埋入孵化介质中进行孵化。一切准备妥帖之后，孵化盒放入孵化器中。

Left: Newly hatched *G. prasina*. Right: Newly hatching *G. boulengeri* (both images Matthew Most).

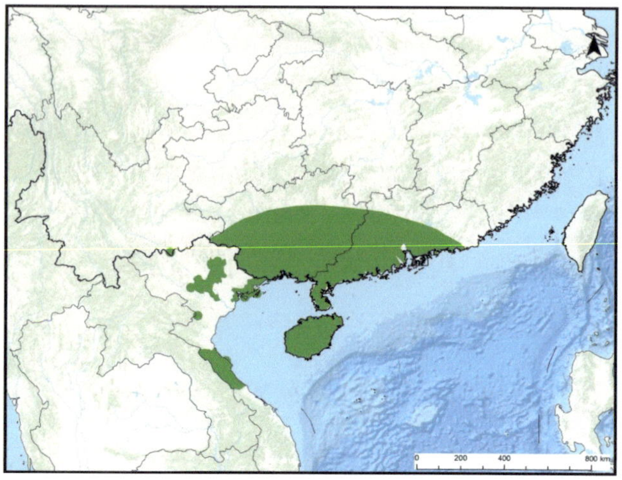

Scientific name / 学名: *Gonyosoma* [formerly *Rhyncophis*] *boulengeri* (Mocquard, 1897)

English name / 英文名: Rhino ratsnake, Rhinoceros snake

Chinese name / 中文名: 尖喙蛇 / jiān huì shé

Etymology / 种名词源: Named after George Albert Boulenger (1858 - 1937), a herpetologist at the British Museum of Natural History, London / 以两栖和爬行动物学家George Albert Boulenger（1858-1937）的名字来命名。他曾在英国自然历史博物馆任职。

Terra typica / 模式产地: Xuy Nong Chao, Ha Long Bay, Gulf of Tonkin, Quang Ninh province, Vietnam (N20.617° E107.167°)/ 越南，广宁省，下龙湾，Xuy Nong Chao

Length / 体长: Maturity at 900 mm; Hatchlings emerge around 170-270 mm. Maximum males 1380 mm (347 mm tail), females 1130 mm (285 mm tail) / 体长900毫米时达到性成熟；刚孵化的幼蛇体长170～270毫米。已知最大雄性个体全长1380毫米，其中尾长347毫米；最大雌性个体全长1130毫米，其中尾长285毫米。

Description: This is a medium-bodied species. The cephalic scales are as follows: supralabials 8-10, 4th to 6th, or 5th to 7th, in contact with orbit; infralabials 10-11; loreal present; preocular 1; postoculars 2-3. Dorsal scale rows smooth, numbering 19:19:15; ventrals 207-216; paired subcaudals 123-138; anal divided. This species has an ontogenetic shift in coloration from juvenile to adult. The dorsal coloration in adult *G. boulengeri* is lime to olive-green with white and black flecking around some scales. The ventral coloration is off-white in both juveniles and adults. Dorsally, juveniles are gray-brown and turn green with age. With increasing maturity, this color shifts from brown to green, potentially reflecting a shift in habitat preference, as shown in other species. Adults and juveniles have a dark stripe stretching along with the upper labial scales from the snout through the eye to the jaw angle. A notable feature of this species is the elongated, pointed, rostral appendage.

Left: Juvenile specimen (Nathan Shepard)

Right: Adult specimen (Matthew Most).

外观描述：体型中等，突出的吻端是本种最显著的特点。鳞被：上唇鳞8（3-3-2、4-3-1式），9（3-3-3、4-3-2式）或10（3-3-4、4-3-3式）；下唇鳞10～11；具颊鳞；眶前鳞1；眶后鳞2～3；背鳞平滑，19-19-15行；腹鳞207～216；尾下鳞123～138对；肛鳞二分。小亚成体与成体在体色上有所差异。成体背面黄绿或橄榄绿色，散有白色或黑色细斑。小亚成体背面为灰褐色，随着年龄的增长颜色逐渐变绿，这或与生存环境的变化有关。成体、小亚成体腹面皆为灰白色。本种沿上唇鳞上缘有一道过眼的黑色纵纹，该纵纹始于吻端止于唇末端。

Distribution: Within Greater China, this species is found in Guangdong, Guangxi, Hainan Island, and Yunnan. Outside of Greater China, it has been recorded in northern Vietnam. The species has been found from 300 to 1500 m.

地理分布：在中国，分布于广东、广西、海南和云南。中国以外，分布于越南北部。垂直分布300～1500米。

Habits and habitats: Arboreal. This species is closely associated with riparian forests and often found close to the water in mountainous regions throughout its range. Both subtropical and tropical lowland and montane forests are known to be inhabited. This species is very docile by nature, and even wild specimens rarely bite when picked up.

习性与栖息地：树栖性。亚热带及热带地区的低地、山地森林都是本种典型的栖息环境。常在水源附近活动。生性温顺，即使野生个体被抓取也极少扑咬。

Above: Juveniles are gray-brown and gradually turn green with age (Matthew Most). Below: Habitat on Hainan Island.

Diet and feeding habits: Primarily feeds on mammals with a preference for rodents in captivity. In the wild, this species is known to eat both birds and mammals. Juveniles are known to feed on fish and tadpoles and maybe an adaptation to their hatchling coloration.

食性与喂养：人工饲养个体主要以哺乳动物为食，尤其偏爱啮齿类。自然环境中，已知捕食鸟类和哺乳类，小亚成体会捕食鱼类和蝌蚪。

Captive care: See the general care at the introduction of this genus. Hatchlings are provided live fish prey items in their water dish to entice feeding and then later weaned to rodents. In captivity, this species has been recorded to have a life span of 15-20 years.

人工饲养：具体情况请参阅本属介绍章节中的“人工饲养概述”版块。可将活鱼投入水盆中引诱幼体进食，之后逐步以啮齿类为主进行投喂。据记载，人工饲养下本种的寿命为15-20年。

Placing a dead, frozen/thawed pinky mouse in with fish is one way to entice juvenile *G. boulengeri* to transition into feeding on mammals. Here is a juvenile *G. boulengeri* feeding on a small minnow (Matthew Most).

Reproduction: Oviparous. Sexual maturity is achieved at 3-4 years. In captivity, specimens are provided with a brumation period of 10-15℃ for 3-4 months. Breeding is observed in the spring following brumation. The gestation period for *G. boulengeri* is 40-60 days, with eggs being laid in June-July. Clutch size ranges from 9-15 eggs with incubation of eggs at 23-25℃ for 60-70 days. Hatchlings have a brown-gray coloration and undergo an ontogenetic change to green with maturity. Females are also capable of producing 1-2 clutches a year. Females and males should be housed together throughout the breeding season as it has been identified that ovulation can be missed by not having a male present for insemination. Males may subdue the female by biting on the back of the neck of the prospective mate.

Some freshly hatched juveniles (Matthew Most).

繁殖：卵生。3～4年达到性成熟。人工饲养下，个体应在10～15℃下经历3～4个月的冬化期。据观察，交配在冬化后的春季进行。母蛇怀卵40～60天。6～7月产卵。一窝卵9～15枚。一年产卵1～2窝。孵化温度23～25℃，孵化期60～70天。幼蛇体色为灰褐色，成年后转为绿色。为提高繁殖成功率，整个繁殖季公母个体应置于同一环境中饲养。交配时，雄性或通过咬住雌性颈背来制服雌性。

Taxonomy and nomenclature: This species was originally in *Elaphe*, then placed in its genus, *Rhynchophis*, most recently placed in *Gonyosoma*, and has been once again proposed to be back in *Rhynchophis*.

分类与命名：本种最初为锦蛇属成员，之后曾改隶尖喙蛇属（*Rhynchophis*），目前则归隶树栖锦蛇属，不过也有学者提议将本种重新归入尖喙蛇属。

Conservation Status: Ranked as Least Concern (LC) by the IUCN Redlist, last assessment 2011.

保护状况：本种在《濒危物种红色名录》中被定为“无危”等级（最近一次评定为2011年）。

A) *Gonyosoma prasinum* / 绿锦蛇

B) *G. frenatum* / 灰腹绿锦蛇

C) *Cyclophiops* [*Ptyas*] *major* / 翠青蛇

D) *Trimeresurus stejnegeri* / 福建竹叶青蛇

E) *Trimeresurus albolabris* / 白唇竹叶青蛇

Similar species: This species shares an overall green coloration as an adult shared with several other species (above), but the presence of the rostral appendage is enough to distinguish this species from any other species of snake in Asia.

相似物种：尖喙蛇成年后通体为绿色，外观上与部分蛇种有一定的相似性。不过突出的吻端足以将本种与亚洲其它蛇种相区分。

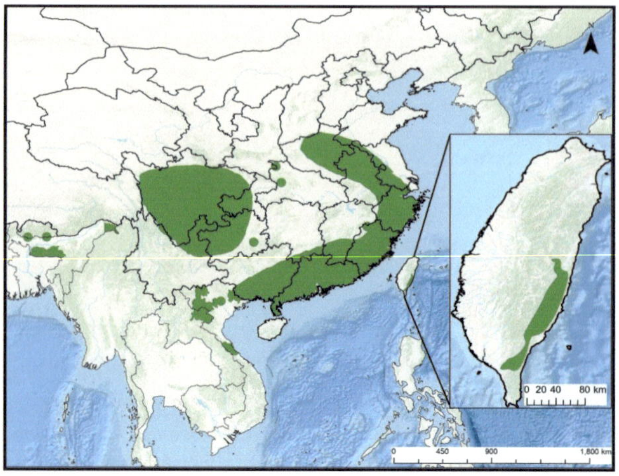

Scientific name / 学名: *Gonyosoma* [formerly *Rhadinophis*] *frenatum* (Gray, 1853)

English name / 英文名: Rein ratsnake, Khasi Hills trinket snake

Chinese name / 中文名: 灰腹绿锦蛇 / huī fù lǜ jǐn shé

Etymology / 种名词源: Lat. *frenum* = meaning bridle, rein, bit. This is likely in reference to the conspicuous "bridle-like" stripe from the snout to the back of the jaws / 源自拉丁语中的"*frenum*"一词，意为"马勒"。该种名或意指灰腹绿锦蛇头背处形似马勒的斑纹。

Terra typica / 模式产地: Khasi Hills, Meghalaya, India, / 印度，梅加拉亚邦，卡西丘陵

Length / 体长: Maturity at 600-650 mm (males), 800 mm (females); Hatchlings emerge at 120-150 mm. Maximum males 1475 mm, 270 mm tail, females 1500 mm, 310 mm tail / 雄性体长600～650毫米时达到性成熟，雌性体长800毫米时达到性成熟。刚孵化的幼蛇体长120～150毫米。已知最大雄性个体全长1475毫米，其中尾长270毫米；最大雌性个体全长1500毫米，其中尾长310毫米。

Description: This is a slender-bodied species. The cephalic scales are as follows: supralabials 8-9, with 3rd to 5th, or 4th to 6th in contact with orbit; infralabials 9-11; preocular 1; postoculars 2; temporals 2+2, 2+3, or 1+2. Loreal absent (fused with prefrontal into a single shield). Ventrals 198-235 with a strong lateral keel; anal divided; subcaudals 118-149, paired. Adults have faintly keeled scales in rows of 19:19 (rarely 17):17 (rarely 15 or 13). Snout (when viewed laterally) is wedge-shaped. Grass or olive green (rarely turquoise) background coloration. Interstitial skin dark blue-black. The upper lip and underside lighter green. Juveniles are gray and turn green with age.

外观描述：体修长。鳞被：上唇鳞8（2-3-3、3-3-2式）或9（2-3-4、3-3-3式）；下唇鳞9～11；眶前鳞1；眶后鳞2；颞鳞2+2、2+3或1+2；无颊鳞（与前额鳞合为一枚鳞片）；腹鳞198～235，侧棱明显；肛鳞二分；尾下鳞118～149对；背鳞19-19（少数17）-17（少数15或13）行，成体背鳞微弱起棱。吻部侧视呈楔形。体背草绿或橄榄绿色（少数蓝绿色）。鳞间皮肤深蓝色。上唇及腹面淡绿色。小亚成体背面灰色，随着年龄的增长体色逐渐变绿。

Specimen from Shennongjia, Hubei (Zhonghua Yang 杨忠华)

Distribution: Within Greater China, the species is found in Anhui, Chongqing, Fujian, Guangdong, Guangxi, Guizhou, Henan, Hubei, Jiangxi, Shaanxi, Sichuan, Taiwan, and Zhejiang. It is unusually absent from the islands of Hong Kong and Hainan. Outside of China, it is also found in India, Laos, Myanmar, and Vietnam. Found as high as 2000 m.

地理分布：在中国，分布于安徽、重庆、福建、广东、广西、贵州、河南、湖北、江西、陕西、四川、台湾和浙江。海南和香港未有分布记录，这点颇令人意外。中国以外，本种分布于印度、老挝、缅甸和越南。垂直分布高达海拔2000米。

Habits and habitats: Poorly studied. Diurnal and highly arboreal. Found in shrubs, bamboo clumps, and trees in mountain forests. When threatened, strikes repeatedly and vibrates tail.

习性与栖息地：此方面研究甚少。白昼活动，高度树栖。常见于山林之中，栖息于灌木、竹子、树木之上。若感知威胁，常抖动尾部并反复发动攻击。

Diet and feeding habits: Primarily feeds on mammals with a preference for rodents in captivity. In the wild, this species is known to eat both birds and mammals. Juveniles are known to feed on fish and tadpoles and maybe an adaptation to their hatchling coloration.

食性与喂养：人工饲养个体主要以哺乳动物为食，尤其偏爱啮齿类。自然环境中，已知猎食鸟类和哺乳类。小亚成体会捕食鱼类和蝌蚪。

Habitat of *G. frenatum* in Wuyishan, Fujian, China.

Captive care: See the general care at the introduction of this genus. Sphagnum moss is used as a substrate for hatchlings as they are very secretive. Hatchlings are provided live fish prey items in their water dish to entice feeding and then later weaned to rodents. In captivity, this species has been recorded to have a life span of 10-15 years.

人工饲养：具体情况请参阅本属介绍章节中的“人工饲养概述”版块。为便于躲藏，幼体的底材建议选用泥炭藓。可将活鱼投入水盆中引诱幼体进食，之后逐步以啮齿类为主进行投喂。据记载，人工饲养下本种的寿命为10～15年。

Reproduction: Oviparous. Sexual maturity is achieved at 3-4 years. In captivity, specimens are provided with a brumation period of 10-15°C for 3-4 months. Breeding is observed in the spring following brumation. The gestation period for *G. frenatum* is 40-60 days, with eggs being laid in June-July. Clutch size ranges from 9-12 eggs with incubation of eggs at 23-25°C for 60-70 days. Hatchlings measure 120-150 mm in total length and have a brown-gray coloration and undergo an ontogenetic change to green with maturity. Males and females should be housed together throughout the breeding season as it has been identified that ovulation can be missed by not having a male present for insemination. This species has only been recorded and observed to lay a single clutch a year.

繁殖：卵生。3～4年达到性成熟。人工饲养下，个体应在10～15℃下经历3～4个月的冬化期。据观察，交配在冬化后的春季进行。母蛇怀卵40～60天。6～7月产卵。一窝卵9～12枚。孵化温度23～25℃，孵化期60～70天。刚孵化的幼蛇体长120～150毫米，体色为灰褐色，成年后转为绿色。为提高繁殖成功率，整个繁殖季公母个体应置于同一环境中饲养。据记载和观察，本种一年只产卵一窝。

Bottom left: Juvenile with gray body coloration (Matthew Most); Top right: A subadult in transition coloration, from Wuyishan in northern Fujian province (Xu Zikun 徐自坤). Bottom right: Adult coloration, specimen from India (Ashok Captain).

Above: *Gonyosoma frenatum* specimen from Guizhou. Below: *G. frenatum* specimen from Henan (both images Gernot Vogel).

Taxonomy and nomenclature: Previously in the genus *Rhadinophis*, recently reassigned to *Gonyosoma*. There are no known subspecies.

分类与命名：之前为绿蛇属（*Rhadinophis*）成员，目前则归隶树栖锦蛇属。本种目前无已知的亚种。

Conservation Status: Not assessed by the IUCN Redlist.

保护状况：本种在《濒危物种红色名录》中被定为“未予评估”等级。

Similar species: Within Greater China, this species could easily be confused with *Gonyosoma prasinum* or *Cyclophiops [Ptyas] major*. To the casual observer, this species could also be confused with *Trimeresurus albolabris* and *T. stejnegeri*.

相似物种：较易与绿锦蛇（*Gonyosoma prasinum*）和翠青蛇（*Cyclophiops major*）相混淆。非专业人士也或将灰腹绿锦蛇同白唇竹叶青蛇（*Trimeresurus albolabris*）和福建竹叶青蛇（*Trimeresurus stejnegeri*）相混淆。

A) *Gonyosoma prasinum*, this species is most easily distinguished by lacking a black postocular stripe/ 区别于绿锦蛇：绿锦蛇头侧无一道过眼的黑色纵纹（Matthew Most）。

B) *Cyclophiops [Ptyas] major*, as with *G. prasinum*, this species is most easily distinguished by lacking a black postocular stripe / 区别于翠青蛇：翠青蛇头侧无一道过眼的黑色纵纹。

C) *Trimeresurus albolabris* is a venomous viper found in southern China. In addition to lacking a black postocular stripe, this species is further distinguished by the presence of a white ventrolateral line in males and a yellow ventral in females, having keeled scales, a loreal pit between the eye and nostril, and by having an elliptical pupil with usually a yellow, orange, or red iris / 区别于白唇竹叶青蛇：白唇竹叶青蛇为一种分布于中国南部的有毒蝰蛇品种，具如下辨识特征：头侧无一道过眼的黑色纵纹。雄性背鳞最外行有一白色纵纹。雌性腹面呈黄色。鳞片起棱。眼睛和鼻孔之间有颊窝。瞳孔呈直立椭圆形，眼虹膜通常为黄色、橘色或红色。

D) *T. stejnegeri* is another venomous viper found throughout central and eastern China. This species can be distinguished from *G. frenatum* by lacking a black postocular stripe, having a white and red ventrolateral line in males and a white ventrolateral line in females, keeled scales, a loreal pit between the eye and nostril, and a red iris with an elliptical pupil / 区别于福建竹叶青蛇：福建竹叶青蛇为一种分布于中国中、东部的有毒蝰蛇品种，具如下辨识特征：头侧无一道过眼的黑色纵纹。雄性背鳞最外行有一红白各半的纵纹，雌性背鳞最外行有一白色纵纹。鳞片起棱。眼睛和鼻孔之间有颊窝。瞳孔呈直立椭圆形，眼虹膜为红色。

Scientific name / 学名: *Gonyosoma* [formerly *Rhadinophis*] *prasinum* (Blyth, 1854).

English name / 英文名: Green ratsnake, Green trinket snake

Chinese name / 中文名: 绿锦蛇 / lǜ jǐn shé

Etymology / 种名词源: Gr. prasinos, meaning "leek-green, light green" / 源自希腊语中的"*prasinos*"一词，意为"韭绿色、淡绿色"。

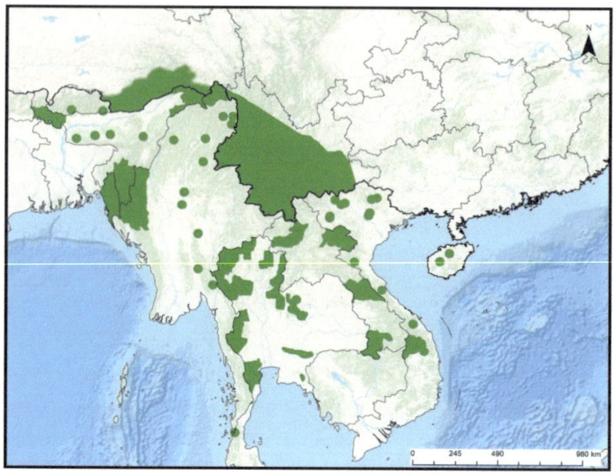

Terra typica / 模式产地: Khasi Hills, eastern Meghalaya State, India (N25.583 E91.633) / 印度，梅加拉亚邦东部，卡西丘陵

Length / 体长: Maturity at 600-650 mm (males), 800 mm (females); Hatchlings emerge around 180-280 mm. Maximum males 915 mm (215 mm tail), females 902 mm (217 mm tail) / 雄性体长600～650毫米时达到性成熟，雌性体长800毫米时达到性成熟。刚孵化的幼蛇体长180～280毫米。已知最大雄性个体全长915毫米，其中尾长215毫米；最大雌性个体全长902毫米，其中尾长217毫米。

Description: The cephalic scales are as follows: Supralabials 9 rarely 8, with 4-6 touching in contact with orbit; preocular 1; loreal present; postoculars 2; temporals 1+2, 2+2, or 1+1. Scales in 19: 17-19: 15-17 rows, with smooth or feebly keeled scales. Ventrals 186-209 with a strong lateral keel; anal single or divided; subcaudals 91-111, paired. Snout rounded. Green body. Upper lip and underside lighter green than upper body. The skin between the scales is black or white. A round pupil helps distinguish it from green pit vipers, all of which have vertical pupils. The iris is often blue or green. By contrast with *G. frenata,* juveniles resemble adults.

Adult *G. prasinum* from central Thailand (Parinya Pawangkhanant).

外观描述： 鳞被：上唇鳞9（3-3-3式），少数8（3-3-2式）；眶前鳞1；具颊鳞；眶后鳞2；颞鳞1+2、2+2或1+1；背鳞平滑或微弱起棱，颈部19行，中段17～19行，肛前15～17行；腹鳞186～209，侧棱明显；肛鳞完整或二分；尾下鳞91～111对。吻圆。体背绿色。上唇及腹面淡绿色。鳞间皮肤黑色或白色。瞳孔圆形，这也是本种与竹叶青蛇属蛇种（*Trimeresurus spp.*）主要的区分依据之一（竹叶青蛇属蛇种的瞳孔呈直立椭圆形）。眼虹膜通常为蓝色或绿色。与灰腹绿锦蛇不同，本种小亚成体与成体在外观上相似。

Distribution: Within Greater China, this species is found in Guizhou, Hainan Island, Sichuan, and Yunnan. Elsewhere: India, Myanmar (= Burma), Thailand, West Malaysia, Laos, Vietnam. Found from 75 to 2650 m.

地理分布： 在中国，分布于贵州、海南、四川和云南。中国以外，分布于印度、缅甸、泰国、马来西亚西部、老挝和越南。垂直分布75～2650米。

Habits and habitats: Diurnal. Primarily arboreal. Found near water in hilly forests.

习性与栖息地： 白昼活动，高度树栖。常见于山林中的水源地带附近。

Generalized habitat shot of *G. prasinum* on Hainan Island.

Diet and feeding habits: Primarily feeds on mammals with a preference for rodents in captivity. In the wild, this species is known to eat both birds and mammals. In captivity, hatchlings have also been known to feed on grasshoppers and other insects. The species kills prey via constriction.

食性与喂养：人工饲养个体主要以哺乳动物为食，尤其偏爱啮齿类。自然环境中，已知捕食鸟类和哺乳类。人工饲养下，幼体也会进食蚱蜢等昆虫。有绞杀猎物的习性。

Captive care: See the general care at the introduction of this genus. In captivity, this species has been recorded to have a life span of 10-15 years.

人工饲养：具体情况请参阅本属介绍章节中的“人工饲养概述”版块。据记载，人工饲养下本种的寿命为10～15年。

Reproduction: Oviparous. Sexual maturity is achieved at 3-4 years. In captivity, specimens are provided with a brumation period of 10-15℃ for 3-4 months. Breeding is observed in the spring following brumation. The gestation period for *G. prasinum* is 50-60 days, with eggs being laid in June-July. Clutch size ranges from 3-14 eggs with incubation of eggs at 23-25℃ for 60-80 days. Females are also capable of producing 1-2 clutches a year. Males and females should be housed together throughout the breeding season as it has been identified that ovulation can be missed by not having a male present for insemination.

繁殖：卵生。3～4年达到性成熟。人工饲养下，个体应在10～15℃下经历3～4个月的冬化期。据观察，交配在冬化后的春季进行。母蛇怀卵50～60天。6～7月产卵。一窝卵3～14枚。孵化温度23～25℃，孵化期60～80天。雌性一年可产卵1～2窝。为提高繁殖成功率，整个繁殖季公母个体应置于同一环境中饲养。

Unlike juvenile *G. boulengeri* or *G. frenatum*, juvenile *G. prasinum* hatch out green in coloration and remain this color throughout their life (Matthew Most).

Top: Specimen from Oudomxay province, Laos (Peter Brakels). Bottom: Head shot of a specimen from Thungyai Naresuan wildlife sanctuary, Kanchanaburi province, Thailand (Parinya Pawangkhanant).

Taxonomy and nomenclature: This species was previously in *Elaphe*, *Rhadinophis*, and most recently placed within the genus *Gonyosoma*. There are no subspecies.

分类与命名：曾归隶锦蛇属和绿蛇属，目前则归于树栖锦蛇属。本种目前无已知的亚种。

Conservation Status: Considered Least Concern (LC) by the IUCN Redlist, last assessment 2011.

保护状况：本种在《濒危物种红色名录》中被定为"无危"等级（最近一次评定为2011年）。

Similar species: This species could easily be confused with *Ptyas [Cyclophiops] major* or *Gonyosoma frenatum*. To the casual observer, this species could also be confused with *Trimeresurus albolabris* and *T. stejnegeri*.

相似物种：较易与翠青蛇（*Cyclophiops major*）和灰腹绿锦蛇相混淆。非专业人士也或将绿锦蛇同白唇竹叶青蛇（*Trimeresurus albolabris*）和福建竹叶青蛇（*Trimeresurus stejnegeri*）相混淆。

Next page / 下一页：

A) *Gonyosoma frenatum*, this species is most easily distinguished by possessing a black post-ocular stripe, which *G. prasinum* lacks / 区别于灰腹绿锦蛇：灰腹绿锦蛇头侧有一道过眼的黑色纵纹，绿锦蛇则无（Ashok Captain）。

B) *Cyclophiops [Ptyas] major* usually has a silver or dark iris, whereas *G. prasinum* has a blue or green iris. *G. prasinum* may have white or black interstitial skin, whereas *C. major* only has black / 区别于翠青蛇：翠青蛇眼虹膜通常为银色或黑色，而绿锦蛇眼虹膜为蓝色或绿色。绿锦蛇鳞间皮肤为白色或黑色，而翠青蛇鳞间皮肤仅为黑色。

C) *Trimeresurus albolabris* is a venomous viper found in southern China. This species is distinguished by the presence of a white ventrolateral line in males and a yellow ventral in females, having keeled scales, a loreal pit between the eye and nostril, and by having an elliptical pupil with usually a yellow, orange, or red iris / 区别于白唇竹叶青蛇：白唇竹叶青蛇为一种分布于中国南部的有毒蝰蛇品种，具如下辨识特征：雄性背鳞最外行有一白色纵纹。雌性腹面呈黄色。鳞片起棱。眼睛和鼻孔之间有颊窝。瞳孔呈直立椭圆形，眼虹膜通常为黄色、橘色或红色。

D) *T. stejnegeri* is another venomous viper found throughout central and eastern China. This species can be distinguished from *G. prasinum* by having a white and red ventrolateral line in males and a white ventrolateral line in females, keeled scales, a loreal pit between the eye and nostril, and a red iris with an elliptical pupil / 区别于福建竹叶青蛇：福建竹叶青蛇为一种分布于中国中、东部的有毒蝰蛇品种，具如下辨识特征：雄性背鳞最外行有一红白各半的纵纹，雌性背鳞最外行有一白色纵纹。鳞片起棱。眼睛和鼻孔之间有颊窝。瞳孔呈直立椭圆形，眼虹膜为红色。

A
B
C
D

The genus *Oreocryptophis* / 紫灰蛇属

This is a monotypic genus.

Etymology: From the Greek words *oros,* meaning mountain, *kryptos*, meaning hidden or secret, and *ophis*, meaning snake. The generic name alludes to the animal's secretive way of life in mountainous areas.

本属为单型属，即属下仅紫灰锦蛇（*Oreocryptophis porphyraceus*）一个物种。

属名词源：源自希腊语，“*oros*”意为“山”，“*kryptos*”意为“隐藏的”，“*ophis*”意为“蛇”。该属名暗示了属下蛇种在山区地带隐秘的生活方式（故该属有时也被称作“山隐蛇属”）。

General captive care / 人工饲养概述：

If available, captive-bred specimens should be acquired. Wild-caught or farm-raised specimens typically have parasites, are malnourished, and are heavily stressed, resulting in a weakened immune system. If working with wild-caught or farm-raised individuals, a qualified veterinarian should conduct a fecal and lung wash to test for parasites. Animals should be quarantined according to the protocols set out by the facility or the keeper, but an approximate quarantine period of 3 months to 1 year is advised to prevent the spread of any potential diseases or parasites.

野外捕获个体及部分养殖场所提供的个体或因寄生虫感染、营养不良、应激反应等因素存在免疫力低下的隐患。若购入以上个体，建议联系有资质的兽医对个体的呼吸道和消化道样本进行寄生虫检测。若可能，应从可靠渠道获取人工繁殖个体。新入手的个体需实行隔离，隔离时长可根据提供方给出的建议。一般建议隔离3个月至1年。

Adult specimens and juveniles should be offered a minimum terrarium of 61 x 41 x 16 cm. Hatchlings can be housed in a plastic shoebox or commercialized snake breeding containers measuring 31 x 19 x 10 cm. Males of *Oreocryptophis* should not be housed together as they will combat the lesser individual to death. Paper towels are used as a substrate for hatchlings, as this genus has a strong feeding response and is used to avoid accidental ingestion of substrate debris. Hatchling setups should also include a moist hide which contains sphagnum moss, which is commonly used when the juveniles are opaque. Individuals can be found buried in the substrate with a temperature range of 21-25°C and relative humidity of 60-70%. These animals cycle with weather patterns in the fall when temperatures gradually cool, stimulating brumation and follicular development in females. These animals are maintained at a temperature of 10-15°C for 3-4 months. The genus is known to lay multiple clutches a year and retain sperm for successive clutches in captivity.

成体及小亚成体的饲养容器应不小于61 x 41 x 16厘米。幼体可饲养在普通塑料盒或专业饲养箱内，空间大小建议在31 x 19 x 10厘米。雄性个体间会互相打斗，故不可群养。考虑到本属蛇种有较强的摄食反应，为避免误食底材，幼体的底材建议选用厨房纸。幼体的环境中应放置加湿躲避穴，可在其中填充泥炭藓用于提高湿度。小亚成体蜕皮时也常使用该类躲避穴。当环境温度达到21～25℃，相对湿度介于60～70%，个体或钻入底材之中。为模拟本属蛇种在自然环境中的节律，饲主应在秋季逐步降低饲养环境的温度以刺激个体的冬化行为以及雌性卵泡的发育。冬化时的温度应保持在10～15℃，为期3～4个月。已知雌性一年产卵数窝，并可将雄性的精液保存在体内较长时间，从而保证一次交配可使数窝卵受精。

Egg Incubation: Eggs of this genus are housed in commercially available incubators set to 25-27°C or room temperature and measuring a minimum of 20 x 20 x 15 cm. Incubating at room temperature will result in a hatching delay by approximately 10-15 days. It is best to set these eggs at lower temperatures as the embryo can fully absorb the yolk sac. While the incubation period maybe longer, the neonates are larger and accept larger prey items. The containers must be well ventilated with holes on all sides to allow for fresh airflow and remove excessive moisture. Excessive moisture and improper ventilation can create issues for developing embryos/eggs, as mold and fungus can grow on the egg's surface. Eggs can be incubated upon a variety of media, but preference is vermiculite and perlite. Both substrates retain moisture and aid in maintaining humidity. Still, both media can dry out and may require additional water if necessary. To achieve proper moisture for incubating, dry media should be measured to have a 5 cm depth within the container. A 2:1 ratio of substrate to water by weight is normally sufficient to ensure proper moisture and humidity during incubation. One can check the appropriate balance by squeezing the mixture by hand. Expressing water means the water content is too high. After achieving the proper moisture, the media, followed by the eggs, can be placed in the incubator. When eggs are laid, it is best to gently separate eggs to allow for adequate airflow within the incubation chamber. Eggs are placed atop the media, and a means of suspending the eggs over the media is used for this genus to avoid undue moisture uptake. As a precaution, eggs should not be buried in the substrate for this genus.

卵的孵化：使用专业孵化器以25～27℃或室温进行孵化皆可使本属蛇种的幼体顺利出壳，但室温下的孵化期要多10～15天。孵化器的尺寸建议不小于20 x 20 x 15厘米。此处推荐以低温进行孵化，这有助于胚胎对卵黄囊充分吸收。孵化期将会延长，但出壳后的幼蛇体型明显更大，这也有利于其进食更大的食物。盛放卵的孵化盒四周均应有孔，以保证空气的流通及多余湿气的排出。湿度过高或通风不畅将对卵的发育造成影响，易令卵壳表面滋生真菌或霉菌。孵化介质有多种选择，推荐使用蛭石或珍珠岩。两者皆有不俗的锁水性，有利于湿度的保持。然而，两种介质也可能在孵化的过程中变得过干，若有必要，需另行补水。为确保孵化过程中湿度始终处于理想水平，孵化介质（干燥状态下）在容器内的铺设厚度需达到5厘米。然后，水和孵化介质在重量上以1:2进行配比。一种检验配比是否合适的有效方法是用手捏兑水后的孵化介质。若有水渗出，则表明水份过高。放置卵时，建议互相之间保持一定的距离，以促进孵化盒内空气的流通。卵可放置在孵化介质表面或悬于孵化介质之上。第二种方法可有效降低因湿度过高对孵化造成的不利影响。切不可将本属蛇种的卵埋入孵化介质中进行孵化。一切准备妥帖之后，孵化盒放入孵化器中。

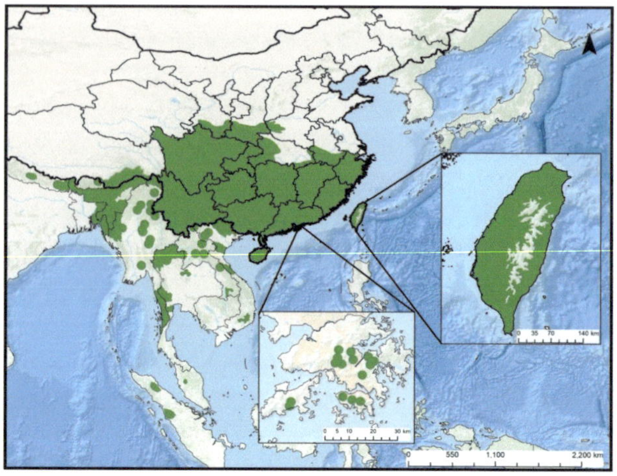

Scientific name / 学名: *Oreocryptophis porphyraceus* (Cantor, 1839)

English name / 英文名: Bamboo ratsnake, trinket snake, red mountain racer, red bamboo trinket snake

Chinese name / 中文名: 紫灰锦蛇 / zǐ huī jǐn shé

Etymology / 种名词源: Derived from Lat. *porphyreticus,* an adjective meaning "made of porphyry" (which is a purple streaked marble), in reference to the coloration of some individuals / 源自拉丁语中的"*porphyreticus*"一词。该词为形容词，意为"由斑岩（一种紫红色的岩石）所构成的"，该种名意指紫灰锦蛇部分个体的体色。

Terra typica / 模式产地: Assam, India / 印度，阿萨姆邦

Length / 体长: Mature size for males: 550-600 mm, for females 650-750 mm. Hatchlings emerge at 170-300 mm. Record size of 1200 mm / 雄性体长550～600毫米时达到性成熟，雌性体长650～750毫米时达到性成熟。刚孵化的幼蛇体长170～300毫米。已知最大个体全长1200毫米。

Description: This is a medium-sized species. The cephalic scales are as follows: supralabials 7-8; infralabials 8-11; preoculars 1; postoculars 2; anterior temporals 0 or 1; posterior temporals 2; scales smooth; dorsal scales are in 17-19 rows at midbody; there are 177-218 ventrals (not angled or keeled) and 47-80 paired subcaudals; the anal plate is divided. The loreal is small, the prefrontal extending well down the side of the face. This species undergoes ontogenetic change from juvenile to adult. Juveniles are usually brightly colored, often yellow with black bands, and become more reddish and duller in coloration with age. In juveniles, there are usually two dorsolateral red stripes, which turn increasing black with age.

An adult (left) from Wuyishan, Fujian and juvenile (right) from Shennongjia, Hubei of *O. p. vaillanti.*

外观描述：体型中等。鳞被：上唇鳞7～8；下唇鳞8～11；眶前鳞1；眶后鳞2；前颞鳞无或1；后颞鳞2；体中段背鳞17～19行，鳞片平滑；腹鳞177～218，无侧棱；尾下鳞47～80对；肛鳞二分；前额鳞较大，弯向头侧；颊鳞较小；小亚成体与成体在外观上有所差异。小亚成体体色通常十分艳丽，体背常为黄色，具黑色横斑。随着年龄的增长，体色渐暗，颜色愈发偏淡红色。小亚成体体背通常具两条红色侧纵线。随着年龄的增长，侧纵线颜色趋于黑色。

Distribution: This species has a very wide distribution but is infrequently observed. Within Greater China, it can be found in Anhui, Chongqing, Fujian, Gansu, Guangdong, Guangxi, Guizhou, Hainan, Henan, Hong Kong, Hubei, Hunan, Jiangsu, Jiangxi, Shaanxi, Sichuan, Taiwan, Tibet, Yunnan, and Zhejiang. Elsewhere, it is found in Bhutan, Cambodia, India, Indonesia, Laos, Malaysia, Myanmar, Nepal, Thailand, and Vietnam.. Found from 1000 to 2600 m.

地理分布：虽分布范围极广，但不易被人发现。在中国，分布于安徽、重庆、福建、甘肃、广东、广西、贵州、海南、河南、香港、湖北、湖南、江苏、江西、陕西、四川、台湾、西藏、云南和浙江。中国以外，分布于不丹、柬埔寨、印度、印度尼西亚、老挝、马来西亚、缅甸、尼泊尔、泰国和越南。垂直分布1000～2600米。

O. porphyraceus in habitat from northeastern Thailand (Parinya Pawangkhanant)

Habit and habitats: In the wild, *O. porphyraceus* inhabits a wide range of habitats from montane forests to grassy plains with rock and bamboo. This is a very secretive snake, which is terrestrial and spends most of its time burrowing under mats of moss, grass, under logs, or rocks. They are mainly active in the early morning and late afternoon hours. When threatened, the species will often coil its tail and hind body, likely to draw attention away from the head. If individuals are picked up, they will readily bite or musk.

When cornered, specimens will frequently curl their tail in an effort to draw attention away from the head. Wuyishan, Fujian.

习性与栖息地：自然环境中，栖息于多种地貌的环境之中，包括山林、草原、竹林等。生性隐秘，地栖性，大部分时间躲藏于苔藓、草丛、枯木和岩石之下。主要于清晨及黄昏活动。若感知威胁，紫灰锦蛇常会卷曲尾部及体后段。这是一种迷惑策略，令天敌误将尾部当作头部。若被抓取，会扑咬并分泌恶臭液体。

Diet and feeding habits: In captivity, the species will take small mammals such as rodents. This species likely preys on small mammals' nests such as rats, mice, voles, and shrews in the wild. This is inferred, as juveniles and adult animals in captivity prefer multiple smaller rodent prey items. Prey is killed via constriction.

食性与喂养：人工饲养个体以啮齿类等小型哺乳动物为食。根据人工饲养下的食性，推测其在自然环境中捕食鼠、鼩鼱等小型哺乳动物的幼崽。本种有绞杀猎物的习性。

Left: An *O. p. pulcher* x *O. p. vaillanti* intergrade specimen of unknown provenance (Matthew Most).

Right: *O. p. porphyraceus* from NE India (Ashok Captain).

Captive care: See the general care at the introduction of this genus. In captivity, this species has been recorded to have a life span of 8-15 years.

人工饲养：具体情况请参阅本属介绍章节中的“人工饲养概述”版块。据记载，人工饲养下本种的寿命为8～15年。

Reproduction: Oviparous. *O. porphyraceus* reaches sexual maturity in 2-3 years. In captivity, specimens are provided with a brumation period of 10-15°C for 3-4 months. Copulation is observed throughout the year; thus, this species appears prolific, having 2-3 clutches a year. The gestation period is 40-50 days, with clutch size ranging from 2-7 eggs, with each egg measuring about 48 mm long. Eggs require incubation at temperatures of 23-25°C for 45-60 days. Males may subdue the female by biting on the back of the neck of the prospective mate. Ontogenetic change from juvenile to adulthood. Juveniles are usually brightly colored, often with shades of red, orange, and yellow, and become duller with age.

繁殖：卵生。2～3年达到性成熟。人工饲养下，个体应在10～15℃下经历3～4个月的冬化期。全年可见交配。雌性个体较为多产，一年产卵2～3窝。母蛇怀卵40～50天。一窝卵2～7枚。卵长径约48毫米。孵化温度23～25℃，孵化期45～60天。交配时，雄性或通过咬住雌性颈背来制服雌性。小亚成体的体色通常十分艳丽，颜色常偏红、橙或黄色。但随着年龄的增长，体色逐渐暗淡。

Top left: Copulation in *O. p. pulcher* (Matthew Most).

Bottom left: Female *O. p. pulcher* with eggs (Matthew Most).

Bottom right: Newly hatched *O. p. pulcher* (Matthew Most).

Taxonomy and nomenclature: The taxonomy of the subspecies complex is one of constant debate and flux. The subspecies of this species have gone through many reassignments and are poorly defined, and many researchers come up with various conclusions on which ones are valid. Historically, up to nine subspecies have been described, with some researchers believing only four to five subspecies as valid, but often disagreeing on which four to five. Due to the confusing and controversial nature of this ongoing debate, until further, comprehensive studies are done, I am listing all subspecies, both valid and invalid, to provide the most transparency on what is occurring with this group.

O. p. porphyraceus; Distribution: China, India, Laos, Myanmar, Nepal, Thailand, Vietnam, and Bhutan. Type locality: Assam, India. It is diagnosed by black stripes on the flank along the posterior half of the body. Ventrals 179-217; subcaudals 52-80; tongue reddish-brown.

O. p. coxi; Distribution: NE Thailand. Type locality: Phuluang, Thailand. Distinctly dorsolateral stripes, lacking cross-bands in adults, which may be present in juveniles. Ventrals 213; subcaudals 62; tongue rosy-pink.

O. p. hainanus; **Some researchers consider this subspecies a synonym of *O. p. vaillanti.* Type locality and distribution: Hainan Island. Similar to *O. p. nigrofasciatus* (another subspecies no longer recognized by most researchers), but distinguished by having more than 70 subcaudals.

O. p. kawakamii; Type locality and distribution: Taiwan. This subspecies is distinguished by the 6th supralabial touching the parietal and having cross-bands eight dorsal scale rows wide.

O. p. laticinctus; Distribution: West Malaysia, Indonesia (Sumatra) and Thailand. Type locality: Cameron Highlands, Pahang, Malaysia. Much wider cross-bands, covering 9-15 rows of dorsal scales. Ventrals 191-201; subcaudals 62-67; tongue reddish-brown.

O. p. nigrofasciatus; **No longer recognized by most researchers; Distribution: Hong Kong, Laos, mainland China, Vietnam. Type locality: Singapore (doubted), Smith (1930) designated E China as the type locality. It is diagnosed by a black flank stripe along the entire body. Ventrals 208; subcaudals 78; tongue color not reported.

O. p. pulcher; Some researchers consider this subspecies a synonym of *O. p. porphyraceus.* It is diagnosed by by fewer ventrals (177-185) and subcaudals (51-56). Adult pattern fades with age in some individuals. Distribution: Southwestern China, Thailand, Myanmar, and Vietnam. Type locality: Kunming, Yunnan, China.

O. p. sikiangensis; **No longer recognized by most researchers. Distribution: Guangdong and Fujian. Type locality: Luofushan, Guangdong, China. Similar to *O. p. nigrofasciatus* (another subspecies no longer recognized by most researchers), ventrals 195; subcaudals 72.

O. p. vaillanti; Distribution: Southern China and northern Vietnam. Type locality: China. Diagnosed by a black head stripe not exceeding the parietal's posterior edge, black stripes on the flank along the whole body. Ventrals 187-209; subcaudals 55-75; tongue reddish-brown.

Conservation status: The conservation status of this species has not been evaluated.

分类与命名：紫灰锦蛇各亚种的定义和分类迄无定论且充满变数。具体划分至今已经历了多次变化。对于究竟哪些为有效亚种，不同学者间的意见并不统一。传统观点认为本种有九大亚种，但不少学者仅认可其中四、五种有效。对于具体哪四、五种有效，分歧依然存在。考虑到个中的复杂性，为尽可能透明公开，本书将对这九大亚种悉数进行介绍。关于各亚种的有效性，进一步的解答仍待未来更全面的研究。

O. p. porphyraceus：分布于中国、印度、老挝、缅甸、尼泊尔、泰国、越南和不丹。模式产地：印度，阿萨姆邦。本亚种可通过体后段的两条黑色侧纵线辨别。腹鳞179～217；尾下鳞52～80对。舌红褐色。

O. p. coxi：分布于泰国东北部。模式产地：泰国，黎府，蒲銮县。侧纵线明显。成体体背无横斑，小亚成体体背或有横斑。腹鳞213；尾下鳞62对。舌玫瑰粉色。

O. p. hainanus：**部分学者认为本亚种为*O. p. vaillanti*亚种的同物异名。模式产地及分布：中国海南。外观上与*O. p. nigrofasciatus*亚种相似（大部分学者已不再认可该亚种的有效性），但可通过尾下鳞超过70对来辨别。

O. p. kawakamii：模式产地及分布：中国台湾。本亚种可通过以下特征辨别：第6枚上唇鳞切顶鳞。体背横斑约占8行鳞宽。

O. p. laticinctus：分布于马来西亚西部、印度尼西亚苏门答腊岛以及泰国。模式产地：马来西亚，彭亨州，金马仑高原。体背横斑更宽，约占9～15行鳞宽。腹鳞191～201；尾下鳞62～67对。舌红褐色。

O. p. nigrofasciatus：**大部分学者已不再认可本亚种的有效性。分布于中国大陆、中国香港、老挝和越南。模式产地：新加坡（存疑）。Smith（1930）将中国东部定为模式产地。本亚种可通过以下特征辨别：两条黑色侧纵线贯穿全身。腹鳞208；尾下鳞78对。关于舌的颜色，尚未见报道。

O. p. pulcher：部分学者认为本亚种为*O. p. porphyraceus*亚种的同物异名。本亚种可通过以下特征辨别：腹鳞数较小（177～185）；尾下鳞51～56对。随着年龄的增长，部分成体体背的斑纹逐渐褪去。分布于中国西南部、泰国、缅甸和越南。模式产地：中国，云南省，昆明市。

O. p. sikiangensis：**大部分学者已不再认可本亚种的有效性。分布于广东和福建。模式产地：中国，广东省，罗浮山。外观上与*O. p. nigrofasciatus*亚种相似（大部分学者已不再认可该亚种的有效性）。腹鳞195；尾下鳞72对。

O. p. vaillanti：分布于中国南部和越南北部。模式产地：中国。本亚种可通过以下特征辨别：头背中央的黑线不过顶鳞后缘。两条黑色侧纵线贯穿全身。腹鳞187～209；尾下鳞55～75对。舌红褐色。

保护状况：本种在《濒危物种红色名录》中被定为“未予评估”等级。

Specimen from northern Taiwan.

North to south variation in the *vaillanti—kawakamii—nigrofasciatus* subspecies complex.

Localities are as follows, A: Fujian, B: Taiwan, C: NE Guangdong, D: southern Hunan, E: NW Guangdong, F: Hong Kong (Dan Rosenberg), G: northern Vietnam (Matthew Most), H: Tao Dao, Vietnam (Rainer Fesser)

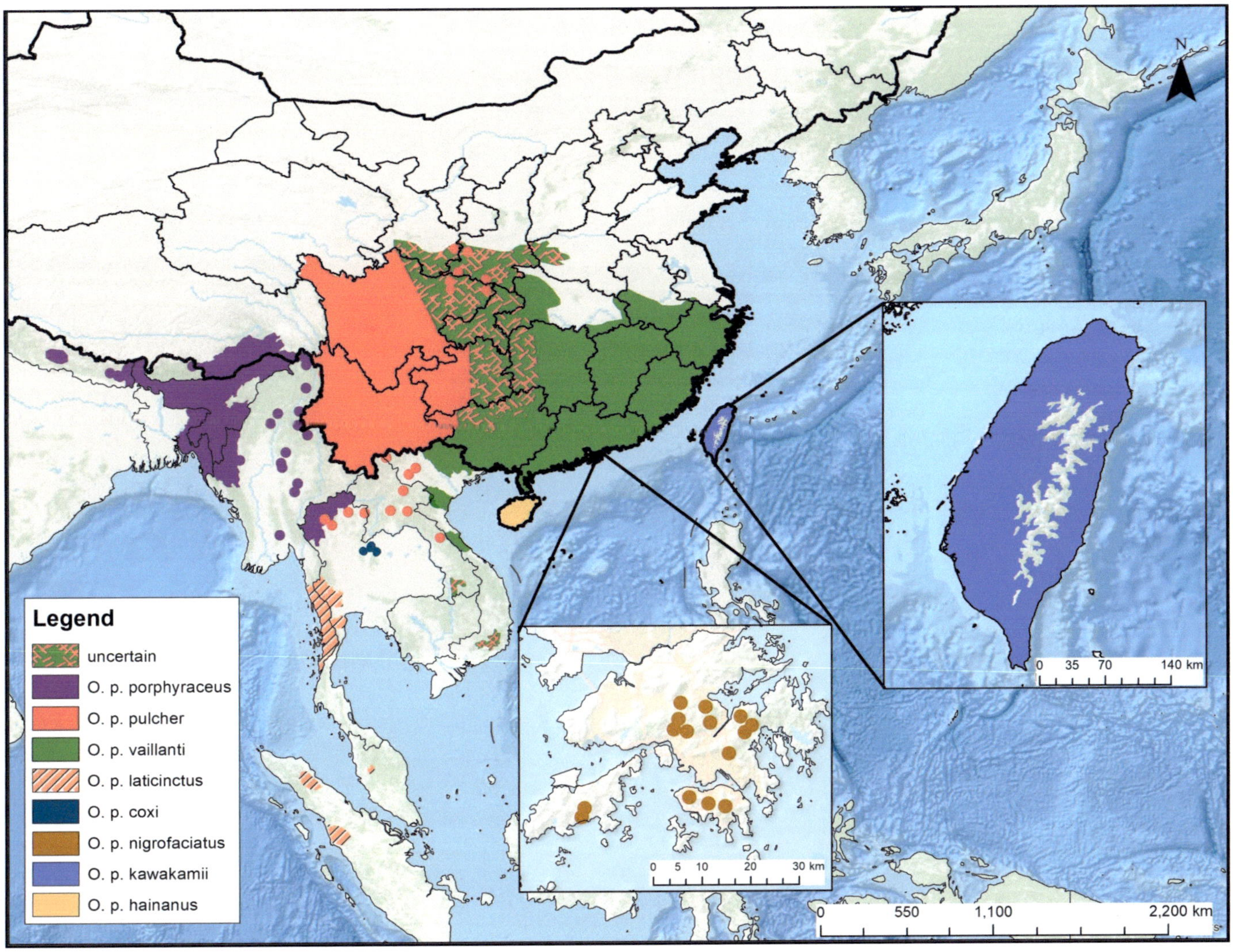

As the designations and definitions of the various subspecies are still unclear, and given that sampling throughout most of the range of the species is also very poor, this distribution map is simply one possible distribution map attempting to delimit the ranges of the various subspecies. For the sake of argument, under this distribution, this map is still recognizing nearly all subspecies, many of which are likely not valid. Further investigations are needed among this highly debated complex of animals before we can say with confidence the precise ranges of the different subspecies. This map is simply this author's first attempt at presenting the subspecies' distributions based on literature sources and images of individuals throughout the species' entire range. As the science becomes more solidified, and we can say with confidence which subspecies are no longer valid, and future iterations of this map should become more accurate.

上图仅是作者根据相关文献和影像资料对各亚种的地理分布做出的首次尝试性描述。由于紫灰锦蛇各亚种的名称和定义仍不清晰，对该物种不同地区个体的采样研究也并不充分，故上图仅代表各亚种可能的地理分布。为便于讨论，上图将所有亚种都涵盖在内，不排除其中某些或为无效亚种，此方面更准确的信息仍待未来进一步的科学研究。

Geographic variation of juvenile *O. porphyraceus* throughout the range (excluding *O. p. coxi*).

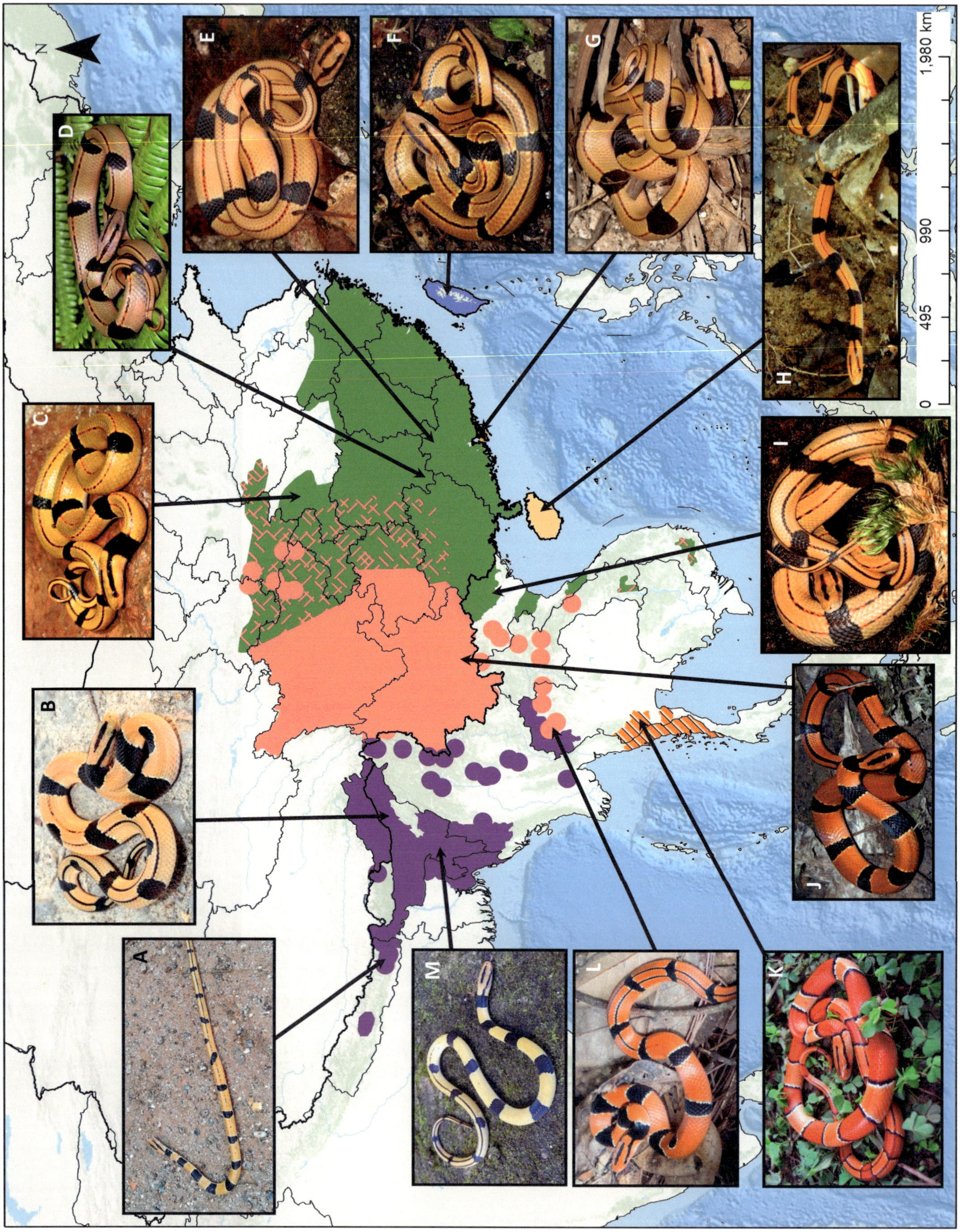

Photo credits: A. Binaya Bhusal; B. Rajib Rudra Tariang; C. author; D. author; E. Dan Rosenberg; F. Elliot Pelling; G. Abdel Bizid; H. Lu Gang 卢刚; I. D. Rosenberg; J. Gernot Vogel; K. Lawan Chanhome ; L. Hein Bijlmakers; M. HT Lalremsanga

Geographic variation of adult *O. porphyraceus* throughout the range (excluding *O. p. coxi*).

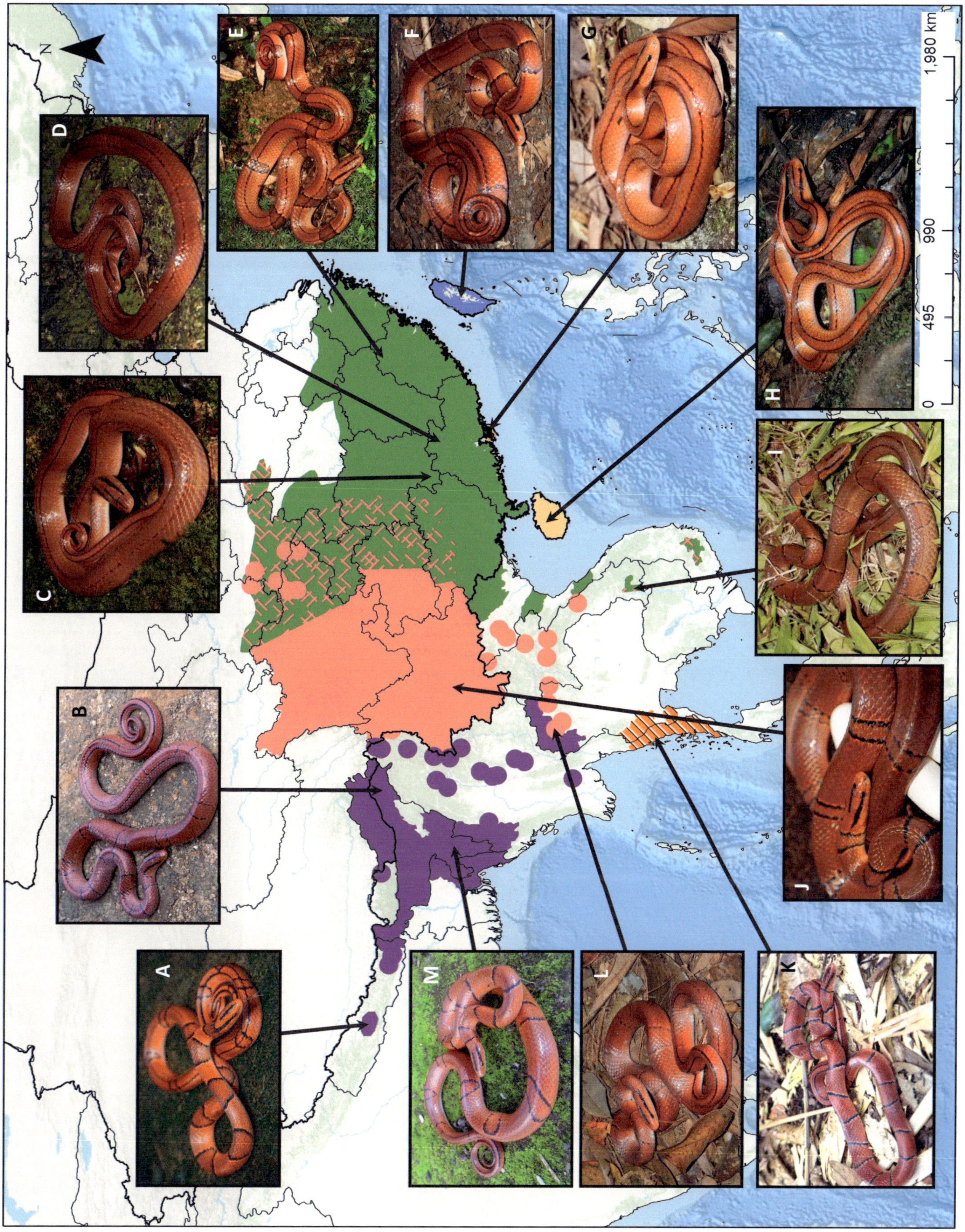

Photo credits: A. Rohit Giri; B. Ashok Captain; C. author; D. author; E. author; F. author; G. Dan Rosenberg; H. Lee Kwok Shing 李国诚; I. Gernot Vogel; J. Matthew Most; K. Mali Naiduangchan; L. Akrachai Aksornneam; M. HT Lalremsanga

Transition series from different regions. Row 1: *O. p. porphyraceus*, specimens from Mizoram, India (HT Lalremsanga); Row 2: *O. p.* cf. *porphyraceus*, specimens from NW Thailand (Hein Bijlmakers, Kamphol Udomritthiruj, Akrachai Aksornneam); Row 3: *O. p. pulcher*, specimens from Yunnan (Matthew Most) and Thailand (Parinya Pawangkhanant); Row4: *O. p. laticinctus*, specimens from Malaysia (M. Most).

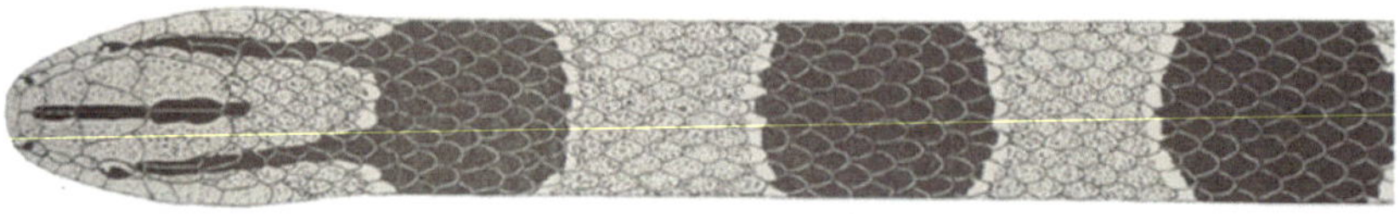

Above: From Schulz (1996): Illustration of juvenile *O. p. porphyraceus* and *O. p. pulcher*.

juvenile / 小亚成体 subadult / 亚成体 adult / 成体

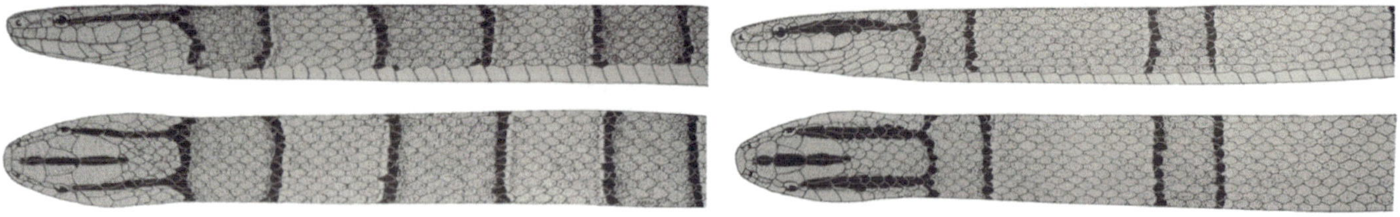

From Schulz (1996): Lateral and dorsal illustrations of adult *O. p. porphyraceus* (left pair) and *O. p. pulcher* (right pair).

Transition series for the *vaillanti—nigrofasciatus—hainanus—kawakamii* subspecies complex. Locations are north to south, top to bottom as follows: Taiwan (Kerry Nelson, Rainer Fesser, K. Messenger), Hunan (Dan Rosenberg, D. Rosenberg, Konrad Mebert), Hong Kong (Abdel Bizid, Lee Wing Ho 李詠豪, D. Rosenberg), Hainan (Lu Gang 卢刚, Lee Kwok Shing 李国诚), Tam Dao Vietnam (R. Fesser, Gernot Vogel, R. Fesser).

juvenile / 小亚成体 | subadult / 亚成体 | adult / 成体

Taiwan

Hunan

Hong Kong

Hainan (middle and last image, same individual)

Tam Dao, Vietnam

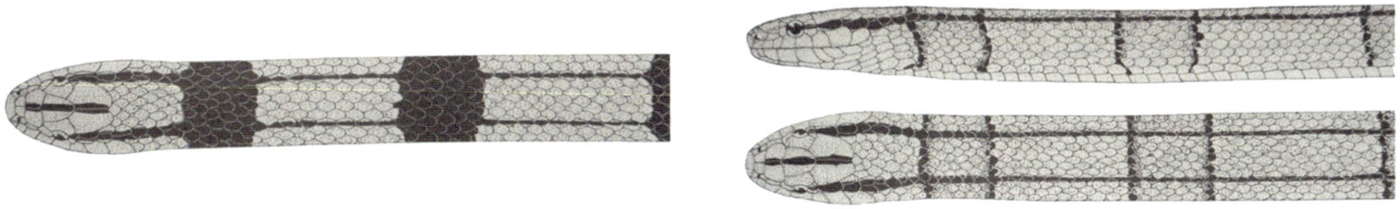

From Schulz (1996): Juvenile (left) and adult (right) illustration of *O. p. vaillanti/ hainanus/ nigrofasciatus/ kawakamii.*

Transition series of juvenile *O. p. coxi* to adult. Unlike all other *O. porphyraceus* subspecies juveniles, juvenile *O. p. coxi* hatch out usually with no cross bands. Occasionally, there will be a few cross bands, but they rarely extend beyond the dorsolateral stripes. As consequence, there is not much of a "transition" from juvenile to adult. If juveniles do have a few vertebral bands between the dorsolateral stripes, the only transition is that these often fade as the individual matures.

O. p. coxi 亚种从小亚成体至成体阶段的变化。同紫灰锦蛇其它亚种不同，*O. p. coxi* 亚种出壳时体背通常无横斑。虽偶有具横斑个体，但横斑在绝大多数情况下限于两条侧纵线之间。随着年龄的增长，这些横斑将逐渐褪去，这也是该亚种从小亚成体至成体阶段唯一的外观变化。

O. p. coxi juvenile (top) and adult (below) (Matthew Most).

Coloration of newly hatched *O. porphyraceus* subspecies. Members represented by A, B. and C lack anterior dorsolateral stripes.

A. Appearance of juveniles of *O. p. porphyraceus* (HT Lalremsanga).

B. Appearance of juveniles of *O. p. pulcher* (Matthew Most).

C. Appearance of juveniles of *O. p. laticinctus* (Konrad Mebert).

D. Appearance of juveniles of *O. p. vaillanti, O. p. kawakamii, O. p. hainana,* and *O. p. nigrofasciatus.* Note the dorsolateral red and black line that runs the length of the body.

E. Appearance of juveniles of *O. p. coxi,* note the lack of cross-bands present in all other members of the species complex (Matthew Most).

Similar species: There are a number of black and yellow colored snakes that could be mistaken for a juvenile *O. porphyraceus*. *O. porphyraceus* is distinct from the majority of these in a number of ways. For one, only juveniles are yellow and black. As they age, they turn more brownish to reddish. As juveniles, their black bands are greatly spaced out (approximately 12—16 along the length), compared to *Lycodon* or *Bungarus* specimens, which have well over 20 black bands.

A. Juvenile *O. porphyraceus*, note that the dorsal bands do not go all the way to the ventrals, but nearly so.
B. *Lycodon flavozonatus*, distinguished by having black saddles instead of black bands.
C. *Lycodon liuchengchaoi*, distinguished by having many more additional black bands, which are of equal width, or greater than the alternating yellow bands, and the bands encircle the entire body. By comparison, on *O. porphyraceus*, the width of the black bands are much more narrow than the alternating yellow bands, which comprise the bulk of the pattern.
D. *Bungarus fasciatus*, this species has a triangular body in cross-section, the tail is blunt, the yellow and black bands are of equal width, encircle the entire body, and the number of black bands is much greater in *B. fasciatus* than in juvenile *O. porphyraceus* (Adam Francis).

In addition to all of these similarities, it should be reiterated that this only applies to juvenile *O. porphyraceus*. If the specimen in question is bigger than 30 cm, and consists of a yellow and black coloration, then *O. porphyraceus* can be ruled out.

相似物种：体色为黄、黑两色的蛇种或被误认为紫灰锦蛇的小亚成体。紫灰锦蛇同它们中的绝大多数蛇种在诸多方面存在着明显差异。例如，只有紫灰锦蛇的小亚成体呈黄、黑两色；随着年龄的增长，体色会愈发偏红棕色。紫灰锦蛇小亚成体全身有12～16个黑色横斑，横斑之间的距离相对较大。而白环蛇属（*Lycodon*）和环蛇属（*Bungarus*）某些蛇种身上的黑色横斑数量超过20个。

A. 紫灰锦蛇小亚成体背面的黑色横斑虽并未完全延伸至腹面，但已十分贴近。

B. 区别于黄链蛇（*Lycodon flavozonatus*）：黄链蛇体背的黑斑呈鞍形。

C. 区别于刘氏白环蛇（*Lycodon liuchengchaoi*）：刘氏白环蛇体背黑色横斑数量更多，各黑斑基本等宽或宽于相间隔的黄色横斑。各横斑环围腹面。相较之下，紫灰锦蛇小亚成体背面的黑色横斑远窄于相间隔的黄色横斑，体背主色为黄色。

D. 区别于金环蛇（*Bungarus fasciatus*）：金环蛇躯干横切面呈三角形，尾末端圆钝。体背黑、黄横斑基本等宽，且皆环围腹面。相较于紫灰锦蛇小亚成体，金环蛇通身的黑斑数量要多得多。

此处有必要重申：仅紫灰锦蛇的小亚成体同以上蛇种有一定的相似性。若某蛇体型大于30厘米且体背呈黄黑两色，该蛇为紫灰锦蛇的可能性可被排除。

Comments: In Taiwan, this species is known locally by a different Chinese name, it is known as: 紅竹蛇 hóng zhú shé.

注释：在中国台湾，本种被称为“紅竹蛇”。

Auliya, Mark, Sandra Altherr, Daniel Ariano-Sanchez, Ernst H. Baard, Carl Brown, Rafe M. Brown, Juan Carlos Cantu, et al. 2016. "Trade in Live Reptiles, Its Impact on Wild Populations, and the Role of the European Market." *Biological Conservation* 204 (July): 103–19.

Beolens, Bo, I Michael Watkins, and I Michael Grayson. 2013. "The Eponym Dictionary of Amphibians." *Choice Reviews Online* 51 (04): 51-1823-51–1823.

Boulenger, George A. 1890. *The Fauna of British India Including Ceylong and Burma. Reptilia and Batrachia.* Edited by W. T. Blanford. 1st ed. London: Taylor & Francis.

Brown, Roland Wilbur. 2000. *Composition of Scientific Words.* Smithsonian Books.

Brunner, M. 1995. "Morphologische Analyse Der Palaarktischen Schlangengruppe Elaphe Dione (Reptilia, Serpentes)." University of Zurich.

Burbrink, Frank T., and Robin Lawson. 2007. "How and When Did Old World Ratsnakes Disperse into the New World?" *Molecular Phylogenetics and Evolution* 43 (1): 173–89.

Chan-ard, Tanya, John W. K. Parr, and Jarujin Nabhitabhata. 2015. *A Field Guide to the Reptiles of Thailand.* 1st ed. New York: Oxford University Press.

Chen, Xin, Ke Jiang, Peng Guo, Song Huang, Dingqi Rao, Li Ding, Hirohiko Takeuchi, et al. 2014. "Assessing Species Boundaries and the Phylogenetic Position of the Rare Szechwan Ratsnake, Euprepiophis Perlaceus (Serpentes: Colubridae), Using Coalescent-Based Methods." *Molecular Phylogenetics and Evolution* 70 (1): 130–36.

Chen, Xin, Alan R. Lemmon, Emily Moriarty Lemmon, R. Alexander Pyron, and Frank T. Burbrink. 2017. "Using Phylogenomics to Understand the Link between Biogeographic Origins and Regional Diversification in Ratsnakes." *Molecular Phylogenetics and Evolution* 111: 206–18.

Cheng, Songlin, Shuyu Wu, and Lin Cheng. 2009. "Diversity of Reptiles in Wuyishan National Nature Reserve of Jiangxi Province." *Jiangxi Science* 27 (6): 827–31.

Cox, Merel J. 1991. *The Snakes of Thailand and Their Husbandry.* Krieger Publishing Co.

Cox, Merel J., Peter Paul van Dijk, Jarujin Nabhitabhata, and Kumthorn Thirakhupt. 1998. *A Photographic Guide to the Snakes and Other Reptiles of Peninsular Malaysia, Singapore and Thailand.* Ralph Curtis Pub.

Dai, Zhongxin, Qiren Yang, Rusong Zhang, Ming Zhang, Dinfu He, Faqin Wu, and Wenzhen Fang. 1996. "Studies on Reptilia Fauna of Hubei Province." *Journal of Central China Normal University Natural Science* 30 (1): 92–95.

Dai, Zong-Xing, Dao-Liang Shi, Shao-Bin Wu, Guo-Yue Liu, Jiao-Fa Shi, Dao-Jin Liu, Dao Sun, and Ru-Sheng Li. 2006. "Preliminary Survey on Amphibia and Reptilia in Wanghu Wetland Nature Reserve in Hubei Province." *Journal of Central China Normal University Natural Science* 40 (2).

Dai, Zongxing, Zhizhang Zheng, Renhu Gong, Hui Huang, Junbo Xiang, Hongmao Zhang, and Xusheng Liu. 2009. "Field Survey on Amphibians and Reptiles in Houhe National Nature Reserve." *Chinese Journal of Zoology* 44 (6): 48–53.

Das, Indraneil. 2012. *A Naturalist's Guide to the Snakes of South-East Asia: Malaysia, Singapore, Thailand, Myanmar, Borneo, Sumatra, Java and Bali.* 1st ed. John Beaufoy Publishing Ltd.

———. 2016. *A Field Guide to the Reptiles of South-East Asia.* Bloomsbury Natural History.

Deng, Xuejian, Yiyun Ye, Bing Wang, and Daode Yang. 1999. "Investigation of Reptiles' Community Structure and Diversity in the Yongzhou Dupangling Nature Reserve." *Sichuan Journal of Zoology* 18 (3): 124–27.

Frith, C. B. 1977. "A Survey of the Snakes of Phuket Island and Adjacent Mainland Areas of Peninsular Thailand." *Natural History Bulletin of the Siam Society* 26: 263–316.

Garzoni, Jean., Geniez, Philippe. 2004. "Elaphe Dione (Pallas, 1773), a Snake Taxon New to the Turkish Herpetofauna." *Herpetozoa* 16 (3/4): 174–75.

Giri, Rohit, Roshan Giri, and Kamal Devkota. 2020. "New District Records of Snakes in Nepal." *Reptiles & Amphibians* 27 (3): 442–43.

Grismer, L. Lee, Thy Neang, Thou Chav, Perry L. Wood, Jamie R. Oaks, Jeremy Holden, Jesse L. Grismer, Thomas R. Szutz, and Timothy M. Youmans. 2008. "Additional Amphibians and Reptiles from the Phnom Samkos Wildlife Sanctuary in Northwestern Cardamom Mountains, Cambodia, with Comments on Their Taxonomy and the Discovery of Three New Species." *Raffles Bulletin of Zoology* 56 (1): 161–75.

Gunther, Albert C. L. G. 1864. *The Reptiles of British India*. London: The Ray Society.

Guo, Ke-Ji, Xiao-Qi Mi, and Xue-Jian Deng. 2007. "Reptiles Resources and Fauna of Shunhuangshan Nature Reserve of Hunan Province." *Central South Forest Inventory and Planning* 26 (4): 47–51.

Guo, Peng, Qin Liu, Edward A. Myers, Shaoying Liu, Yan Xu, Yang Liu, and Yuezhao Wang. 2012. "Evaluation of the Validity of the Ratsnake Subspecies Elaphe Carinata Deqenensis (Serpent: Colubridae)." *Asian Herpetological Research* 3 (3): 219–26.

Hecht, Vera L., Cuong T. Pham, Tao T. Nguyen, Truong Q. Nguyen, Michael Bonkowski, and Thomas Ziegler. 2013. "First Report on the Herpetofauna of Tay Yen Tu Nature Reserve, Northeastern Vietnam." *Biodiversityy Journal* 44 (4): 507–52.

Helfenberger, N., and B. Schatti. 1998. "Morphological Adaptations for Egg-Eating in the Snake Elaphe Davidi Sauvage 1884." *Russian Journal of Herpetology* 5 (1): 36–42.

Hobcroft, Dion, and Klaus-dieter Schulz. 2010. "First Record of Orthriophis Cantoris (Boulenger, 1894) in Bhutan." *Sauria* 32 (2): 56.

Hofmann, Sylvia, Peter Fritzsche, and Georg Miehe. 2016. "Correspondence A New Record of Elaphe Dione from High Altitude in Western Sichuan Reveals High Intraspecific Differentiation" 52 (October): 273–77.

Huang, Christina. 2019. "Rhadinophis Frenatus (Gray, 1853) Found in Shaanxi Province, China." *Chinese Journal of Zoology* 54 (5): 771–72.

Huang, Song, Li Ding, Frank T. Burbrink, Jun Yang, Jietang Huang, Chen Ling, Xin Chen, and Yaping Zhang. 2012. "A New Species of the Genus Elaphe (Squamata: Colubridae) from Zoige County, Sichuan, China." *Asian Herpetological Research* 3 (1): 38–45.

Jacobs, Jan-cor. 2005. "Elaphe Dione, the Steppe Rat Snake." *Litteratura Serpentium* 25 (1): 6–11.

Javed, S.M. Maqsood, Farida Tampal, and C. Srinivasulu. 2010. "First Record of Coelognathus Radiatus (Boie, 1827) (Reptilia: Colubridae) from the Ananthagiri Hills, Eastern Ghats, India." *Journal of Threatened Taxa* 2 (9): 1172–74.

Jiang, Zhigang, Jianping Jiang, Yuezhao Wang, E Zhang, Yanyun Zhang, Lili Li, Feng Xie, et al. 2016. "Red List of China's Vertebrates." *Biodiversity Science* 24 (5): 500–551.

Kane, Daniel, Iri Gill, Luke Harding, Joe Capon, Martin Franklin, Francesca Servini, Benjamin Tapley, and Christopher J. Michaels. 2017. "Captive Husbandry and Breeding of Gonyosoma Boulengeri." *Herpetological Bulletin*, no. 139: 7–11.

Khan, Muhammad Sharif. 2002. *Guide to the Snakes of Pakistan*. Frankfurt: Edition Chimaira.

Langeveld, C.M. 1995. "The Chinese Leopard Snake, Elaphe Bimaculata (K.P. Schmidt, 1925) an Easy Snake to Keep." *Litteratura Serpentium* 15 (4): 82–88.

Leong, T. M., and Kelvin K.P. Lim. 2003. "Herpetofaunal Records from Fraser's Hill, Peninsular Malaysia, with Larval Descriptions of Limnonectes Nitidus and Theloderma Asperum (Amphibia: Ranidae and Rhacophoridae)." *Raffles Bulletin of Zoology* 51 (1): 123–36.

Li, Cao, Tao Wen, and Peng Guo. 2009. "A New Snake Record in Sichuan Province - Maculophis Bellus." *Chinese Journal of Zoology* 44 (2): 135–37.

Li, Zushi, Jiwei Mo, Yingle Gu, Song Liu, Dongbo Fei, and Daode Yang. 2010. "Investigation and Analysis on Amphibian and Reptile Resources in Nanyue Hengshan National Nature Reserve." *Hunan Forestry Science and Technology* 37 (1): 20–29.

Liao, Qingyi, Cuncun Yang, Zujie Kang, Guiqing Yu, Jia Tang, Linzhuang Bai, Daode Yang, and Taifu Huang. 2020. "Investigation of Herpetological Diversity in Hunan Hupingshan National Nature Reserve." *Hunan Forestry Science and Technology* 47 (2): 90–97.

Ling, Chen, Shaoying Liu, Song Huang, Frank T. Burbrink, Peng Guo, Zhiyu Sun, and Jie Zhao. 2010. "Phylogenetic Analyses Reveal a Unique Species of Elaphe (Serpentes, Colubridae) New to Science." *Asian Herpetological Research* 1 (2): 1–7.

Liu, Song, and Jieqiong Duan. 2017. "Amphibians and Reptiles Resources in Paiyashan National Forest Park in Hunan Porvince." *Hunan Forestry Science and Technology* 44 (4): 52–63.

Luu, Vinh Quang, Truong Quang Nguyen, Cuong The Pham, Kien Ngoc Dang, Thanh Ngoc Vu, Sladjana Miskovic, Michael Bonkowski, and Thomas Ziegler. 2013. "No End in Sight? Further New Records of Amphibians and Reptiles from Phong Nha – Ke Bang National Park, Quang Binh Province, Vietnam." *Biodiversity Journal* 4 (2): 285–300.

Mell, R. (1922) Beiträge zur Fauna Sinica. I. Die vertebraten Südchinas, Feldlisten und Feldnoten der Säuger, Vögel, Reptilien, Batrachier. *Archiv für Naturgeschichte*, Berlin 88A(10), 1–146, pl. 4.

Nemes, Liza, Randall Babb, Wayne Van Devender, Khoi Vu Nguyen, Quyet Khac Le, Thanh Ngoc Vu, Anna Rauhaus, Truong Q Nguyen, and Thomas Ziegler. 2013. "First Contribution to the Reptile Fauna of Quang Ngai Pro-Vince,Central Vietnam." *Biodiversity Journal* 4 (2): 301–26.

Nguyen, Luan Thanh, Daniel Kane, Manh Van Le, Thang Tai Nguyen, Ha Van Hoang, Timothy E.M. McCormack, Benjamin Tapley, and Sang Ngoc Nguyen. 2020. "The Southernmost Distribution of the Rhinoceros Snake, Gonyosoma Boulengeri (Mocquard, 1897) (Reptile, Squamata, Colubridae), in Vietnam." *Check List* 16 (2): 337–42.

Nguyen, Tan Van, Peter Brakels, Nathanael Maury, Somchit Sudavanh, Parinya Pawangkhanant, and Nikolay A Poyarkov. 2020. "New Herpetofaunal Observations from Laos Based on Photo Records" 14 (2): 218–49.

Nishikawa, Kanto, and Frank T Burbrink. 2014. "The Phylogenetic Position and Taxonomic Status of the Rainbow Tree Snake Gonyophis Margaritatus (Peters, 1871) (Squamata: Colubridae)" 3881 (6): 2012–15.

Orlov, Nikolai L., Sergei A. Ryabov, Tao Thien Truong Quang Nguyen, and Tao Thien Truong Quang Nguyen. 2011. "Rediscovery and Redescription of Two Rare Snake Species: Oligodon Lacroixi Angel et Bourret, 1933 and Maculophis Bellus Chapaensis (Bourret, 1934) (Squamata: Ophidia: Colubridae) from Fansipan Mountains, Northern Vietnam." *Russian Journal of Herpetology* 17 (4): 310–22.

Platt, Steven G., George R. Zug, Kalyar Platt, Win Ko Ko, Khin Myo Myo, Me Me Soe, Tint Lwin, et al. 2018. "Field Records of Turtles, Snakes and Lizards in Myanmar (2009-2017) with Natural History Observations and Notes on Folk Herpetological Knowledge." *Natural History Bulletin of the Siam Society* 63 (1): 67–114.

Pope, Clifford H. 1935. *The Reptiles of China*. 1st ed. Chicago: The American Museum of Natural History.

Rovatsos, M., B. Augstenova, M. Altmanova, M. Sloboda, P. Kodym, and L. Kratochvil. 2018. "Triploid Colubrid Snake Provides Insight into the Mechanism of Sex Determination in Advanced Snakes." *Sexual Development* 12: 251–55.

Sang, Nguyen Van, Ho Thy Cuc, and Nguyen Quang Truong. 2009. "Herpetofauna of Vietnam." *Chimaira, Frankfurt*, 768.

Schleich, Hermann H., and Werner Kastle, eds. 2002. *Amphibians and Reptiles of Nepal*. 1st ed. Ruggell: A.R.G. Gantner Verlag Kommanditgesellsch.

Schmidt, C. 2000. "Morphologie Der Hemipenes von Kletternattern (Elaphe Fitzinger, 1833)." University of Zurich.

Schulz, Klaus-dieter. (1986a) Die hinterasiatischen Kletternattern der Gattung *Elaphe*. Teil IV. *Elaphe dione* (Pallas, 1773). *Sauria*, Berlin 8(1), 27–30.

———. (1986b) Die hinterasiatischen Kletternattern der Gattung *Elaphe*. Teil V. *Elaphe bimaculata* (Schmidt, 1925). *Sauria*, Berlin 8(2), 23–26.

———. 1996. *A Monograph of the Colubrid Snakes of the Genus Elaphe, Fitzinger*. Koeltz Scientific Books.

———. 2013. *Old World Ratsnakes. A Collection of Papers*. Edited by Klaus-dieter Schulz. Bargain Smart Plug.

Schulz, Klaus-dieter, Wolfgang Böhme, and Frank Tillack. 2011. "HEMIPENIS MORPHOLOGY OF Coronella Bella STANLEY , 1917 WITH COMMENTS ON TAXONOMIC AND NOMENCLATURAL ISSUES OF RATSNAKES (SQUAMATA: COLUBRIDAE: COLUBRINAE: Elaphe AUCT .)." *Russian Journal of Herpetology* 18 (4): 273–83.

Schulz, Klaus-dieter, Notker Helfenberger, and Basundhara Chettri. 2010. "Psammophis Nigrofasciatus CANTOR , 1839 , Ein Juniorsynonym von Oreocryptophis Porphyraceus Porphyraceus (CANTOR, 1839)." *Sauria* 32 (3): 31–38.

Sharma, R. C. 2007. Fauna of India and the adjacent countries - Reptilia: 1-410. Zoological Survey of India Kolkata.

Shi, Jingsong, Zhongwen Jiang, and Wei Zhao. 2019. "Ruoergai Snakes Were Discovered in Diebu County, Gansu." *Chinese Journal of Zoology* 54 (5): 769–70.

Shu, Shi, Zongxing Dai, Faqing Wu, Luo Zukui, Qianqian Wang, Zhaoquan Zhu, Cao Guobin, and Yunhai Pu. 2007. "Investigating on Resources of Amphibians and Reptiles in Mountain Wanchaoshan Nature Reserve, Hubei Province." *Journal of Huazhong Normal University Natural Science* 41 (2): 278–81.

Silva, Anslem De, and Kanishka Ukuwela. 2018. *A Naturalist's Guide to the Reptiles of Sri Lanka.* John Beaufoy Publishing Ltd.

Simonov, Evgeniy, Artem Lisachov, Natalia Oreshkova, and Konstantin Krutovsky. 2018. "The Mitogenome of Elaphe Bimaculata (Reptilia: Colubridae) Has Never Been Published: A Case with the Complete Mitochondrial Genome of E. Dione." *Acta Herpetologica* 13 (2): 185–89.

Singh, Abhishek, and Ritesh Joshi. 2018. "Sighting of the Himalayan Trinket Snake, Orthriophis Hodgsonii Günther (1860) (Reptilia: Colubridae), in Sahastra Dhara, Uttarakhand: A New Elevational Record." *Amphibian and Reptile Conservation* 12 (1): 15–17.

Smith, Malcolm A. 1943. *The Fauna of British India Ceylong and Burma, Including the Whole of the Indo-Chinese Sub-Region. Taylor and Francis.* Vol. 3. London: Taylor & Francis.

Song, Lin. 2016. "Archelphae Bella in Wuyishan Jiangxi Province." *Sichuan Journal of Zoology* 35 (1): 334500.

Square, Berjaya Times, and Jalan Imbi. 2010. "Death Feigning Behavior in Three Colubrid" 17 (1): 1–5.

Stanley, Arthur. 1917. "Two New Species of Chinese Snakes." *Journal of the North-China Branch of the Royal Asiatic Society* 47: 83–84.

Staszko, Ray, and Jerry G. Walls. 1994. *Rat Snakes: A Hobbyist's Guide to Elaphe and Kin.* 1st ed. Tfh Publications Inc.

Stejneger, Leonhard. 1996. *Herpetology of Japan and Adjacent Territory.* Society for the Study of Amphibians and Reptiles.

Stuart, Bryan L., and Harold F. Heatwole. 2008. "Country Records of Snakes from Laos." *Hamadryad* 33 (1): 97–106.

Stuebing, Robert B., and Robert F. Inger. 1999. *A Field Guide to the Snakes of Borneo.* Natural History Publications.

Su, HJ. 2019. "Archelaphe Bella Found in Mount Fanjingshan , Guizhou Province." *Chinese Journal of Zoology* 54 (1): 149–50.

Teynié, A., Nguyen, T.Q., Lorvelec, O., Piquet, A., Lottier, A., David, P. 2014. "Amphibiens et Reptiles Du Laos: Nouvelles Données Nationales et Provinciales." *Bull. Soc. Herp. Fr.* 151 (January): 21–52.

Teynié, Alexandre, Anne Lottier, and Patrick David. 2017. "Azemiops Feae Boulenger, 1888 (Squamata: Viperidae) et Cinq Autres Additions Notables Pour La Faune Des Serpents Du Laos." *Bulletin de La Société Herpétologique de France* 1888 (162): 39–54.

Traupman, John C. 1994. *Latin Dictionary.* Second edi. New York: Amsco School Publications, Inc.

Tuniyev, Boris S., Nikolai L. Orlov, Natalia B. Ananjeva, and Aram L Aghasyan. 2019. *Snakes of the Caucasus: Taxonomic Diversity, Distribution, Conservation.* Edited by Natalia B. Ananjeva. Moscow: KMK Scientific Press Ltd.

Uetz, P. (editor), The Reptile Database, http://www.reptile-database.org, accessed 15 Oct 2020.

Utiger, Urs, Nothler Helfenberger, Beat Schätti, Catherine Schmidt, Markus Ruf, and Vincent Ziswiler. 2002. "Molecular Systematics and Phylogeny of Old and New World Ratsnakes, Elaphe AUCT., and Related Genera (Reptilia, Squamata, Colubridae)." *Russian Journal of Herpetology* 9 (2): 105–24.

Utiger, Urs, Beat Schatti, and Notker Helfenberger. 2005. "The Oriental Colubrine Genus Coelognathus Fitzinger, 1843 and Classification of Old and New World Racers and Ratsnakes (Reptilia, Squamata, Colubridae, Colubrinae)." *Russian Journal of Herpetology* 12 (1): 39–60.

Wall, Frank. 1921. "Notes on Some Notable Additions to the Bombay N.H.S. Snakes Collection." *Journal of Bombay Natural History Society* 28: 43–44.

Wallach, Van, Kenneth L. Williams, and Jeff Boundy. 2014. *Snakes of the World, A Catalogue of Living and Extinct Species.*

Wang, Kai, Jinlong Ren, Hongman Chen, Zhitong Lyu, Xianguang Guo Ke Jiang, Jinmin Chen, Jiatang Li, Peng Guo, Yingyong Wang, Jing Che 2020. The updated checklists of amphibians and reptiles of China. Biodiversity Science 28 (2): 189-218.

Wang, Lijun, Zhensheng Liu, Hao Zhai, Jun Han, Qunxiu Liu, and Xiaoming Wang. 2010. "Species Diversity and Faunal Characteristics of Amphibians and Reptiles in Helan Mountains." *Chinese Journal of Ecology* 29 (11): 2293–97.

Wang, Ping, Lei Shi, and Peng Guo. 2019. "Morphology-Based Intraspecific Taxonomy of Oreocryptophis Porphyraceus (Cantor, 1839) in Mainland China (Serpentes: Colubridae)." *Zoological Research* 40 (4): 324–30.

Whitaker, Romulus, and Ashok Captain. 2004. *Snakes of India The Field Guide.* Chennai: Draco Books.

Wogan, Guinevere O U, Jens V Vindum, Jeffery A Wilkinson, Michelle S Koo, Joseph B Slowinski, Htun Win, Thin Thin, et al. 2008. "NEW COUNTRY RECORDS AND RANGE EXTENSIONS FOR MYANMAR AMPHIBIANS AND REPTILES." *Hamadryad* 33 (1): 83–96.

Wu, Tao, Zhixiao Liu, Youxiang Zhang, Benzhong Wang, Zhengxiang Mao, and Taifu Huang. 2019. "Species Diversity and Faunal Characteristics of Amphibians and Reptiles in the Xiaoxi National Nature Reserve, Hunan." *Sichuan Journal of Zoology* 38 (4): 452–58.

Xie, Li, Xianhua Song, Hong Jiang, Zhijun Ma, and Jiang Zhou. 2013. "Elaphe Moellendorffi Was Discovered in Libo County, Guizhou Province." *Chinese Journal of Zoology* 48 (3): 487–89.

Yang, Qiren, Ming Zhang, Dingfu He, Zhongxin Dai, Faching Wu, Rusong Zhang, Weimin Lu, et al. 1998. "Investigated into Species Diversity of the Terrestrial Vertebrates of Nature Protected Area in Houhe, Hubei." *Journal of Central China Normal University Natural Science* 32 (4): 476–77.

Ye, Qiang. 1979. "Brief Introduction of Medicinal Snakes in Our Country," 32–37.

Zeng, Ling-Bo, Shi Shu, Qian-Qian Wang, Fa-Qing Wu, and Zong-Xing Dai. 2008. "Preliminary Survey on Amphibian and Reptilian Resources of Wudaoxia Nature Reserve in Hubei Province." *Journal of Hubei University Natural Science* 30 (4): 407–10.

Zhao, Ermi. 2006. *Snakes of China.* Edited by Er-mi Zhao. 1st ed. Hefei: Anhui Science and Technology Publishing House.

Zhao, Ermi, and Kraig K Adler. 1993. *Herpetology of China.* SSAR & CSSAR.

Ziegler, Thomas, Ralf Hendrix, Vu Ngoc Thanh, Martina Vogt, Bernhard Forster, and Dang Ngoc Kien. 2007. "The Diversity of a Snake Community in a Karst Forest Ecosystem in the Central Truong Son, Vietnam, with an Identification Key." *Zootaxa* 40 (1493): 1–40.

Ziegler, Thomas, and Quang Nguyen. 2010. "New Discoveries of Amphibiens and Reptiles from Vietnam." *Bonn Zoological Bulletin.* 57 (2): 137–47.

Ziegler, Thomas, Dao Thi Anh Tran, Truong Quang Nguyen, Ronith Gila Bina Perl, Lea Wirk, Magdalena Kulisch, Tanja Lehmann, et al. 2014a. "New Amphibian and Reptile Records from Ha Giang Province, Northern Vietnam." *Herpetology Notes* 7 (September): 185–201.

———. 2014b. "New Amphibian and Reptile Records from Ha Giang Province, Northern Vietnam." *Herpetology Notes* 7 (April): 185–201.

Zipkin, Elise F., Graziella V. DiRenzo, Julie M. Ray, Sam Rossman, and Karen R. Lips. 2020. "Tropical Snake Diversity Collapses after Widespread Amphibian Loss." *Science* 367 (6479): 814–16.

Numbers in **bold** refer to pages with photographs.

About the author / 关于作者

Kevin Messenger is an American herpetologist from North Carolina. He has been interested in herpetology from a young age, encouraged by his father, who shares the same love for the field, and his mother, who shares a fondness for the outdoors (but not so much for snakes). He studied zoology under Dr. Harold Heatwole at North Carolina State University. He then continued to get his Master's degree under the guidance of Dr. Thomas Pauley at Marshall University, West Virginia. After graduating from NC State University, Kevin traveled to central China to conduct the first herp survey in the Shennongjia National Park. It was this experience that caused Kevin to fall in love with China. His experience opened his eyes to the great need for more herpetological work in China. This thought caused Kevin to pursue a Ph. D. with the specific interest of working in China. Ultimately, he obtained one Ph. D. from Alabama A & M University and a Ph.D. from Nanjing Forestry University. He is currently a Professor at Nanjing Forestry University, where he established the Herpetology and Applied Conservation Laboratory.

凯文•梅辛杰是一位来自于美国北卡罗来纳州的两栖和爬行动物学家。受父亲的影响，他自年幼就对两栖和爬行动物学产生了浓厚的兴趣。母亲虽对蛇类并不感兴趣，但也是一名户外运动爱好者。在北卡罗来纳州立大学本科学习期间，凯文师从Harold Heatwole博士，主修动物学。随后，他师从Thomas Pauley博士，于西弗吉尼亚州的马歇尔大学获得硕士学位。从北卡罗来纳州立大学毕业后，凯文来到了位于中国中部的神农架国家公园，深入当地进行首次两栖和爬行动物调研工作。正是这段经历使他爱上了中国，同时也让他意识到中国需要加强在两栖和爬行动物领域的相关研究工作。这一想法激励凯文选择继续深造，并促使他重返中国。最终，他分别于阿拉巴马农业机械大学和南京林业大学获得博士学位。目前，凯文任教于南京林业大学，并在校内成功创立了“两栖和爬行动物学与应用保护实验室”。

Made in United States
Cleveland, OH
08 November 2026

47841302R00114